Springer Theses

Recognizing Outstanding Ph.D. Research

For further volumes:
http://www.springer.com/series/8790

Aims and Scope

The series "Springer Theses" brings together a selection of the very best Ph.D. theses from around the world and across the physical sciences. Nominated and endorsed by two recognized specialists, each published volume has been selected for its scientific excellence and the high impact of its contents for the pertinent field of research. For greater accessibility to non-specialists, the published versions include an extended introduction, as well as a foreword by the student's supervisor explaining the special relevance of the work for the field. As a whole, the series will provide a valuable resource both for newcomers to the research fields described, and for other scientists seeking detailed background information on special questions. Finally, it provides an accredited documentation of the valuable contributions made by today's younger generation of scientists.

Theses are accepted into the series by invited nomination only and must fulfill all of the following criteria

- They must be written in good English.
- The topic of should fall within the confines of Chemistry, Physics and related interdisciplinary fields such as Materials, Nanoscience, Chemical Engineering, Complex Systems and Biophysics.
- The work reported in the thesis must represent a significant scientific advance.
- If the thesis includes previously published material, permission to reproduce this must be gained from the respective copyright holder.
- They must have been examined and passed during the 12 months prior to nomination.
- Each thesis should include a foreword by the supervisor outlining the significance of its content.
- The theses should have a clearly defined structure including an introduction accessible to scientists not expert in that particular field.

Suranjana Sengupta

Characterization of Terahertz Emission from High Resistivity Fe-doped Bulk $Ga_{0.69}In_{0.31}As$ Based Photoconducting Antennas

Doctoral Thesis accepted by
Rensselaer Polytechnic Institute, Troy, USA

 Springer

Author
Dr. Suranjana Sengupta
Intel Corporation
Hillsboro
USA
e-mail: suranjana.sengupta@intel.com;
        suranjana@gmail.com

Supervisor
Dr. Ingrid Wilke
Department of Physics, Applied Physics,
  and Astronomy
Rensselaer Polytechnic Institute
110, 8th Street
Troy, NY 12180
USA
e-mail: wilkei@rpi.edu

ISSN 2190-5053

ISBN 978-1-4614-2827-5

e-ISSN 2190-5061

ISBN 978-1-4419-8198-1 (eBook)

DOI 10.1007/978-1-4419-8198-1

Springer New York Dordrecht Heidelberg London

Cover design: eStudio Calamar, Berlin/Figueres

Printed on acid-free paper

Springer is part of Springer Science+Business Media (www.springer.com)

Supervisor's Foreword

The topic of this book is the development and characterization of a terahertz radiation source. On the electromagnetic spectrum, terahertz waves are located between microwaves and the infrared. Terahertz science and technology is attracting great interest due to the development of improved terahertz radiation sources and terahertz radiation detectors. Commonly, terahertz radiation is employed for spectroscopy and imaging. Terahertz spectroscopy measures the dielectric properties of liquid and solid, inorganic and organic materials in the terahertz frequency range. Furthermore, terahertz spectroscopy is a method for vibrational and rotational spectroscopy of gases. Terahertz radiation is invisible and non-ionizing. Water absorbs terahertz radiation strongly. Metals reflect terahertz radiation highly. Materials which are dry or electrically insulating transmit to some extent terahertz radiation. Examples are paper, cardboard, cloths, plastics, oils or air. The capability of terahertz radiation to penetrate a wide range of materials that are opaque to visible light is exploited for terahertz imaging. Terahertz imaging is tested in process and quality control in manufacturing of aircrafts, cars, drugs and computer chips. Furthermore, terahertz imaging is considered for security applications such as screening for concealed weapons.

This book describes the development and characterization of a photoconducting terahertz emitter antenna. The objective of the research is a high-power terahertz radiation source for time-domain terahertz spectroscopy and imaging systems. The research is motivated by na need for photoconducting terahertz emitter antenna for the next generation of multi-watt mode-locked femtosecond lasers. The research presented in this book is a multidisciplinary project at the interface of physics, electrical engineering and materials science.

This book is Suranjana Sengupta's PhD thesis. She performed her doctoral research at the Department of Physics, Applied Physics and Astronomy at Rensselaer Polytechnic Institute. Founded in 1824, Rensselaer Polytechnic Institute is the oldest science and technological university in America. The university was established by Stephen van Rensselaer "for the purpose of instructing persons who may choose to apply themselves in the application of science to the common purpose of life". Rensselaer's vision is exemplified by the research presented in this book.

Suranjana Sengupta decided to study physics while being a high school student. Her choice was motivated by the diverse applications of physics in the frontier area of science and technology. She passed the fiercely competitive joint entrance examination to the prestigious Indian Institute of Technology (IIT) and completed a 5-year Bachelor and Master of Science program in physics at the IIT campus in Kharagpur. She was admitted to the PhD program in physics at Rensselaer Polytechnic Institute in 2004. I met her first when she enrolled in my course on Methods in Theoretical Physics. Later, I accepted her as a graduate research assistant in my research group on Ultrafast and Terahertz Spectroscopy.

Successful experimental research builds on skillful measurements and on deep knowledge of the physical laws relevant to the research topic. The research on terahertz radiation presented in this book is guided by electromagnetic theory and principles of electron transport in semiconductors. The measurement methods are ultrafast near-infrared pump-probe measurements and time-domain terahertz emission measurements. The experimental methods require the alignment of near-infrared laser and invisible terahertz radiation with spatial and temporal precision of less than half a millimeter. The initial learning of the experimental skills requires perseverance of the student.

Today's research requires teamwork. Therefore, we would like to acknowledge gallium indium arsenide crystal growth by our collaborator Partha S. Dutta and his co-workers in the Department of Electrical, Computer and Systems Engineering at Rensselaer Polytechnic Institute. Furthermore, we collaborated with Youngkok Ko on the research of terahertz radiation emission from undoped gallium indium arsenide. Ricardo Ascazubi built the time-domain terahertz emission spectrometer. The experiments involving an Ytterbium laser where conducted in collaboration with Evgueni Slobodtchikov and David Cook. Finally, we would like to acknowledge the enthusiasm and financial support for our research from QPeak, Inc., and Physical Sciences, Inc. Part of the material presented in this book is based upon work supported by the National Science Foundation under Grants Nos. 0619499 and 1002040.

Troy, New York, 17 October 2010   Ingrid Wilke

Acknowledgment

This dissertation remains woefully incomplete without thanking all the people who have made my 5 years of stay at RPI an enjoyable experience and helped me make it this far.

First of all, I wish to express my deepest gratitude to my adviser, Professor Ingrid Wilke, for all her support, guidance, and encouragement that made the course of my studies at RPI a thoroughly enjoyable learning experience. I learned a lot from her during the course of my PhD and I greatly appreciate the privilege of working with her.

I would also like to thank my Doctoral Dissertation Committee members, Professor Partha Dutta, Professor Morris Washington, Professor Kim Lewis, and Professor Peter Persans, for their valuable feedback and review of my doctoral dissertation.

I am thankful to my collaborators for their excellent contributions without which this dissertation would not have been possible. In this regards, I would like to thank QPeak Incorporated for developing the high power Yb:KYW laser, and Dr. David Cook from Physical Sciences Corporation for his valuable inputs on terahertz emission experiments on as-grown $Ga_xIn_{1-x}As$ samples. I am deeply grateful to Professor Partha Dutta for growing all the $Ga_xIn_{1-x}As$ samples used in this dissertation and for the insightful and indepth discussions on sample growth techniques. I would also like to thank Professor Dutta's graduate student Adam Gennet for polishing the semiconductor samples.

I am grateful to Professor X.-C. Zhang for allowing me to use the facilities in his laboratory for running my experiments.

Finally, no words are sufficient to express my gratitude towards my friends and family. But for their endless support and unconditional love I would not have the courage to undertake this journey. Thank you all for being there for me always.

Contents

Chapter 1
Introduction

The generation and detection of terahertz (THz) frequency electromagnetic radiation and the study of materials interaction occurring in this frequency regime has been of considerable interest to the scientific community of late. The term terahertz is typically used to indicate the region of electromagnetic spectrum between the frequencies 100 GHz (100×10^9 Hz) and 10 THz (10×10^{12} Hz) corresponding to the sub-millimeter wavelength range 3 mm to 30 μm between the microwave and the infra-red bands. Terahertz radiation is often commonly referred to as T-rays or simply abbreviated as THz. Much of the scientific interest in T-rays is due to the unique properties of this type of radiation. Unlike X-rays, THz waves have very low photon energy and thus cannot lead to harmful photoionization in biological samples. THz waves are also transparent to most dry dielectric materials like wood, paper, cloth, and plastic and as such suffer less scattering than visible and IR waves due to their longer wavelengths. Furthermore, many biological and chemical compounds exhibit characteristic absorption and dispersion signatures in the THz regime due to vibrational and rotational transitions. This implies that THz radiation might be used to examine the chemical composition of such compounds. Together, these properties make T-rays an excellent source for medical diagnostics and non destructive evaluation type of application. Yet, until late 1980s this part of the electromagnetic spectrum was least explored due to the technical difficulties involved in developing efficient and compact THz sources and detectors.

Over the past decade, advances in photonics and nano-technology have made significant contribution towards the development of more intense THz sources and higher sensitivity detectors [1]. This has enabled the scientific community to exploit the unique properties of THz radiation in a wide variety of applications such as genetics, pharmacy, medical science, industrial non-destructive evaluation, environmental monitoring, security, and basic science [2]. There are numerous examples in every field: genetic diagnostics [3], skin cancer diagnosis [4], classification of polymorphs in medical drugs [5], large scale integrated (LSI) circuit testing [6], narcotics [7] and explosive inspection [8], to name a few. In addition to

S. Sengupta, *Characterization of Terahertz Emission from High Resistivity Fe-doped Bulk Ga$_{0.69}$In$_{0.31}$As Based Photoconducting Antennas*, Springer Theses, DOI: 10.1007/978-1-4419-8198-1_1, © Springer Science+Business Media, LLC 2011

the above, THz imaging is being used as a routine inspection method for the non-destructive evaluation of space shuttle foam to detect defects such as voids in heat insulation pads [9]. In basic science as in industrial application, the contribution of THz technology has been invaluable. THz technology has been used, along with optical index analysis, to probe low energy carrier dynamics in various electronic materials such as superconductors [10].

THz time-domain spectroscopy (TDS) provides a new method to characterize the compositional properties of various types of solids, liquids and gases and even flames and flows. TDS typically relies on a broad band short pulse THz source. The optical generation of THz radiation from such sources falls into two categories. The first involves the generation of ultrafast photocurrent in a photoconductive switch or semiconductor by exciting its surface with a femtosecond (fs) laser pulse whose photon energy is higher than the band gap of the semiconductor material. The second process involves generation of THz pulses by a nonlinear optical process such as optical rectification in materials such as GaSe [11, 12], ZnTe [13, 14], GaP [15], CdTe [16], DAST [17, 18] LiTaO$_3$ [17, 19], and LiNbO$_3$ [19, 20]. In both of these instances the generated THz power is in the nanowatt range. Another intriguing optical THz generation method that has recently attracted considerable attention is ambient air-plasma generation [2, 21]. Here, an intensely pulsed laser creates air-plasma which in turn generates THz radiation. However, this approach is currently far from being practical for applications due to the large laser systems that are required to drive such strong THz-emission.

1.1 Identification and Significance of the Problem

In a typical THz TDS set up [22], an ultrafast near-infrared (NIR) laser pulse illuminates a photoconducting antenna formed by adhering electrodes over a highly resistive semiconductor. When a bias voltage is applied, and an ultrafast pulse having photon energy higher than the band gap of the semiconductor is incident between the electrodes, the resulting photocurrent surge leads to the emission of a single cycle broadband THz pulse. The rise time and bandwidth of the emitted THz pulse is limited by the laser pulse duration and carrier mobility inside the semiconductor, with a typical output spectrum containing components from below 100 GHz up to 3 THz. Ultrafast mode-locked Titanium-Sapphire (Ti:S) lasers delivering 800 nm pulses at 82 MHz repetition rates with average output power of 500 mW are ideally suitable for use with emitters based on semiconductor materials having band gap less than 1.5 eV. The pulse duration of a Ti:S laser system can be as short as tens of femtosecond. However, the THz power obtained from a Ti:S laser pumped time domain THz spectroscopy system is usually less than 1 μW. In many applications such as inspection, screening, and illicit material detection, such low average power results in long signal acquisition time. Low THz power is also a crucial parameter in sub-surface imaging and spectroscopic application where the time-domain THz beam is strongly attenuated due to absorption and scattering in the material. Some THz application can benefit from the use of the amplified femtosecond lasers. Such systems can deliver a few mJ of laser pulse energy but at a low repetition rate (~ 1 kHz) which is a significant drawback for a number of applications.

In spite of the spectacular technical advancements in the field of THz science and technology in the past decade, it is the lack of compact, low-cost, turnkey time-domain THz systems and high-power, high-bandwidth and high repetition rate THz sources that has limited the widespread commercial adoption of THz technology. Enhancing output power and shrinking the size remain some of the challenges for future THz TDS systems. One solution lies in utilizing compact mode locked multi-watt Ytterbium (Yb) based lasers to drive semiconductor based photoconducting (PC) antennas for potentially stronger THz emission. Since the photon energy of the Yb-laser system is well below the band gap of the conventional GaAs based PC antenna, it is imperative to search for alternative narrow band gap semiconductor materials. The ultimate goal is the development of Yb-laser driven sub-picosecond THz radiation source with an average power of 1 mW operating at MHz repetition rates. However, it is first necessary to identify a suitable semiconductor material for this purpose and to carefully characterize the THz emission process from the same to understand its capabilities and limitations. Thus, the goal of this research will be to investigate optically excited THz emission from novel photoconducting antennas based on narrow band gap semiconductor materials and the dependence of the process on semiconductor properties. It can be expected that the knowledge gained through this research will yield valuable insights towards the complete understanding of the THz emission mechanism and can be utilized to optimize the design parameters to develop compact high-power, high-bandwidth and high repetition rate sources of THz radiation.

The recent development of directly diode-pumped high-power laser systems can make a significant contribution towards the improvement of time domain THz TDS systems. While traditional Ti:S based laser systems have found widespread use in scientific studies, they are expensive and limited in oscillator power for generation of THz radiation. This is because the Ti-crystal absorbs in the visible spectral range (514–532 nm) and cannot be directly pumped by near-infrared (808, 980 nm) high-power diode lasers but requires frequency-doubled solid state lasers based on neodymium-doped gain media as the pumping source. The schematics of a diode-pumped mode-locked Ti:S laser is presented in Fig. 1.1a. A neodymium yttrium vanadate (Nd:YVO$_4$) laser crystal is end-pumped by a fiber-coupled diode bar (FCbar) producing about 10 W of 1,064 nm NIR power with a conversion efficiency of about $\sim 40\%$. Frequency doubling converts the 10 W, 1,064 nm NIR light to 5 W of green 532 nm light that becomes the output of the pump laser. Thus, we see that frequency doubling not only adds to the cost, complexity and footprint of the laser system, but also limits the pump power available to the Ti:S oscillator. The overall conversion efficiency of such a diode-pumped mode locked Ti:S laser is only about 2.5% as depicted in Fig. 1.1a.

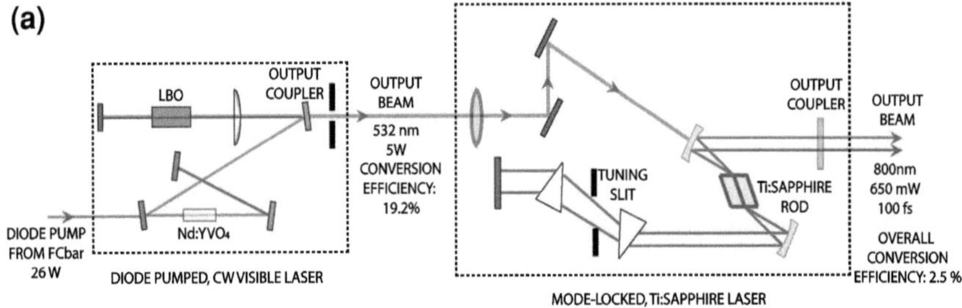

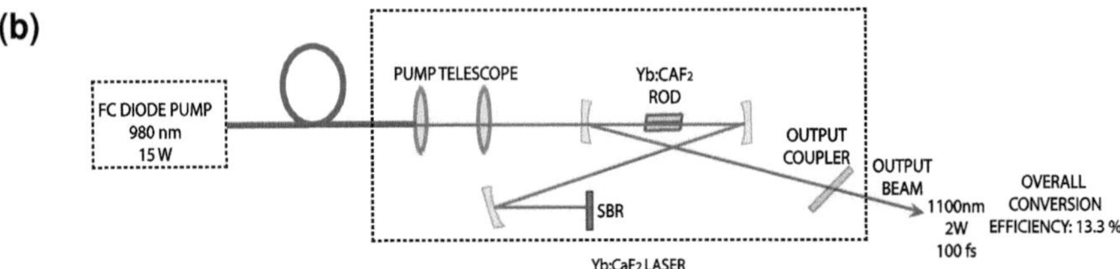

Fig. 1.1 Schematic diagram of **a** Ti:Sapphire laser [25] and **b** Yb:CaF2 [24] lasers depicting overall conversion efficiencies

Recently a large family of Yb-doped bulk laser materials have been developed that can be directly diode-pumped with very high efficiency [23, 24]. Some of these materials are broadly tunable and can potentially support the generation of high-power laser pulses shorter than 100 fs. The schematic of such a laser is depicted in Fig. 1.1b. Since frequency doubling is not required, the overall conversion efficiency in such lasers is much higher ($\sim 13.3\%$ for the present configuration). However, such Yb doped lasers have an emission wavelength of ~ 1.1 μm with photon energies much lower than the band gap of the conventional GaAs based THz emitters and novel narrow band gap THz emitter materials need to be developed.

1.2 Scope of Dissertation

Recent advances in the development of large aperture dc-biased THz emitters [26] have substantially increased the efficiency of conversion of the power available directly from ultrafast mode-locked laser into THz radiation. Large aperture in this context means that the transverse dimensions of the antenna are large relative to the center wavelength of the emitted radiation (1 THz = 300 μm). The most commonly used material for this purpose is low-temperature-grown (LT) GaAs [27]. As mentioned earlier, since the photon energy of the Yb-doped laser is below the GaAs band gap, it is necessary to explore alternative semiconductors. $Ga_xIn_{1-x}As$ is a promising material in this regard since its band gap can be tuned from 3.44 to 0.87 μm by variation of Ga mole fraction from $x = 0$ to $x = 1$ [28]. Earlier research on THz emission from unbiased $Ga_xIn_{1-x}As$ [$0 < x < 0.65$] bulk crystals excited at 800 nm, reports that THz emission varies as a function of material composition [29]. The emission was maximized for Ga mole fraction $x = 0.3$. Pulsed THz emission from externally biased $Ga_{0.47}In_{0.53}As$ and $Ga_{0.7}In_{0.3}As$ has also been successfully demonstrated for excitation wavelengths of 1.55 and 1.06 μm, respectively [30, 31]. However, since ternary semiconductors such as $Ga_xIn_{1-x}As$ are extremely difficult to grow as bulk crystals, all previous research on $Ga_xIn_{1-x}As$ based photoconductive antennas have been restricted to only a few compositions of $Ga_xIn_{1-x}As$ thin films grown by molecular beam epitaxy (MBE) on binary substrates. Furthermore, since the bulk resistivity of as-grown $Ga_xIn_{1-x}As$ is low [$\sim 20 \ \Omega$ cm], THz emission from externally biased large aperture $Ga_xIn_{1-x}As$ based emitters, which operate at high dc voltages have not been reported to date.

The primary goal of this research is to investigate the THz emission characteristics of voltage biased semi-large aperture THz emitters based on high-resistivity $Ga_{0.69}In_{0.31}As$ bulk crystals obtained through novel growth techniques. The term "semi-large" is used to characterize PC antennas with electrode gap spacing between 0.1 and 1 mm. The $Ga_{0.69}In_{0.31}As$ bulk crystals employed in our experiments have been grown by the unique combination of vertical and horizontal gradient freezing methods along with the accelerated crucible rotation technique and exhibit high resistivity, high mobility and femtosecond carrier lifetimes that are much desired for THz PC antennas. The crystals obtained through the above techniques are also expected to have superior electrical and optical properties due to the near-equilibrium growth process compared to thin films grown from vapor or liquid phase. The experimental work in this dissertation primarily focuses on characterizing the THz emission from semi-large aperture voltage biased $Ga_{0.69}In_{0.31}As$ emitters on the basis of (a) material parameters (such as ultrafast mobility and carrier lifetime), (b) external bias applied, and (c) electrode gap spacing, and aims to understand how these parameters might be optimized to develop a high power THz radiation source.

The subsequent sections of this dissertation are developed in the following fashion. Section 2.1 in Chap. 2 provides a review of past research on THz emission from $Ga_xIn_{1-x}As$ based THz emitters. This is followed by Sects. 2.2 and 2.3 presenting a theoretical model of THz emission from PC antennas and semiconductor surface emitters, respectively. In Chap. 3, the principle of transient photo-reflection is described. Chapter 4 deals with the experimental arrangements employed in this research. The details of THz TDS, THz power measurement setup, and ultrafast photo-reflection setups are explained in Sects. 4.1, 4.2 and 4.3, respectively. Section 4.4 describes semiconductor sample growth and preparation techniques. Chapter 5 discusses in detail the results of ultrafast photo-reflection experiments and THz radiation emission from voltage biased $Ga_{0.69}In_{0.31}As$ emitters. The dissertation concludes with Chap. 6 summarizing the results and presenting the future outlook for extending the research on high power THz sources based on voltage biased $Ga_xIn_{1-x}As$ emitters.

References

1. Lee, Y.-S.: Principles of Terahertz Science and Technology, Chaps. 3 & 4. Springer Science+Business Media LLC, New York (2009)
2. Zhang, X.-C., Xu, J.: Introduction to THz Wave Photonics. Springer Science+Business Media LLC, New York (2010)
3. Nagel, M., Richter, F., Brucherseifer, M., Bolivar, P.H., Kurz, H., Bosserhoff, A., Buttner, R.: Integrated THz technology for label free genetic diagnostics. Appl. Phys. Lett. **80**, 154 (2002)
4. Woodward, R.M., Cole, B.E., Wallace, V.P., Pye, R.J., Arnone, D.D., Linfield, E.H., Pepper, M.: Terahertz pulse imaging in reflection geometry of human skin cancer and skin tissue. Phys. Med. Biol. **47**, 3853 (2002)
5. Taday, P.F., Bradley, I.V., Arnone, D.D., Pepper, M.: Using terahertz pulse spectroscopy to study the crystalline structure of a drug: a case study of the polymorphs of ranitidine hydrochloride. J. Pharm. Sci. **92**, 831 (2003)
6. Yamashita, M., Kawase, K., Otani, C., Kiwa, T., Tonouchi, M.: Imaging of large-scale integrated circuits using laser terahertz emission microscopy. Opt. Express **13**, 115 (2005)
7. Shen, Y., Lo, T., Taday, P.F., Cole, B.E., Tribe, W.R., Kemp, M.C.: Detection and identification of explosives using terahertz pulsed spectroscopic imaging. Appl. Phys. Lett. **86**, 1–241116 (2005)
8. Kawase, K., Ogawa, Y., Watanabe, Y., Inoue, H.: Non-destructive terahertz imaging of illicit drugs using spectral fingerprints. Opt. Express **11**, 2549 (2003)
9. Zhong, H., Xu, J., Xie, X., Yuan, T., Reightler, R., Madaras, E., Zhang, X.-C.: Nondestructive defect identification with terahertz time-of-flight tomography. IEEE Sens. J. **5**, 203 (2005)
10. Orenstein, J., Corson, J., Oh, S., Eckstein, J.N.: Superconducting fluctuations in $Bi_2Sr_2Ca_{1-x}Dy_xCu_2O_{8+\delta}$ as seen by terahertz spectroscopy. Ann. Phys. **15**, 596 (2006)
11. Huber, R., Brodschelm, A., Tauser, F., Leitenstorfer, A.: Generation and field resolved detection of femtosecond electromagnetic pulses tunable up to 41 THz. Appl. Phys. Lett. **76**, 3191 (2000)
12. Shi, W., Ding, Y.J., Fernelius, N., Vodopyanov, K.: Efficient, tunable and coherent 0.18–5.27-THz source based on GaSe crystal. Opt. Lett. **27**, 1454 (2002)
13. Nahata, A., Weling, A.S., Heinz, T.F.: A wideband coherent terahertz spectroscopy system using optical rectification and electrooptic sampling. Appl. Phys. Lett. **69**, 2321 (1996)
14. Han, P.Y., Zhang, X.-C.: Free-space coherent broadband terahertz time-domain spectroscopy. Meas. Sci. Technol. **12**, 1747 (2001)
15. Chang, G., Divin, C.J., Hung, L.C., Williamson, S.L., Galvanauskas, A., Norris, T.B.: Power scalable compact THz system based on an ultrafast Yb doped fiber amplifier. Opt. Express **14**, 7909 (2006)
16. Xie, X., Xu, J., Zhang, X.-C.: Terahertz wave generation and detection from a CdTe crystal characterized by different excitation wavelengths. Opt. Lett. **31**, 978 (2006)
17. Hu, B.B., Zhang, X.-C., Auston, D.H., Smith, P.R.: Free-space radiation from electrooptic crystals. Appl. Phys. Lett. **56**, 506 (1990)
18. Zhang, X.-C., Ma, X.F., Jin, Y., Lu, T.-M., Boden, E.P., Phelps, P.D., Stewart, K.R., Yakymyshyn, C.P.: Terahertz optical rectification from a nonlinear organic crystal. Appl. Phys. Lett. **61**, 3080 (1992)
19. Zhang, X.C., Jin, Y., Ma, X.F.: Coherent measurement of terahertz optical rectification from electrooptic crystals. Appl. Phys. Lett **61**, 2764 (1992)
20. Yang, K.H., Richards, P.L., Shen, Y.R.: Generation of far-infrared radiation by picosecond light pulses in LiNbO3. Appl. Phys. Lett. **19**, 320 (1971)
21. Zhong, H., Karpowicz, N., Zhang, X.-C.: Terahertz emission profile from laser-induced air plasma. Appl. Phys. Lett. **88**, 1–261103 (2006)
22. Wilke, I., Sengupta, S.: Nonlinear Optical Techniques for Terahertz Pulse Generation and Detection—Optical Rectification and Electrooptic Sampling. In: Dexheimer, S.L. (ed.)

Terahertz Spectroscopy: Principles and Applications, Optical Science and Engineering, vol. 131, p. 41. CRC Press, Boca Raton (2007)

23. Lucca, A., Debourg, G., Jacquemet, M., Druon, F., Balembois, F., Georges, P., Camy, P., Doualan, J.L., Moncorgé, R.: High-power diode pumped Yb:CaF$_2$ femtosecond laser. Opt. Lett. **29**, 2767 (2004)

24. Druon, F., Chénais, S., Raybaut, P., Balembois, F., Georges, P., Gaumé, R., Aka, G., Viana, B., Mohr, S., Kopf, D.: Diode-pumped Yb:BOYS femtosecond laser. Opt. Lett. **27**, 197 (2002)

25. Tsunami, Spectra-Physics, Manual

26. Planken, P.C.M., van Rijmenam, C.E.W.M., Schouten, R.N.: Opto-electronic pulsed THz systems. Semicond. Sci. Technol. **28**, S121 (2005)

27. Kono, S., Tani, M., Sakai, K.: Generation and detection of broadband pulsed terahertz radiation. In: Sakai, K. (ed.) Terahertz Optoelectronics. Topics in Applied Physics, vol. 97, p. 31. Springer, Berlin (2005)

28. Goldberg, Y.A., Shmidt, N.M.: Gallium indium arsenide. In: Levinshtein, M., Rumayantsev, S., Shur, M.S. (eds.) Handbook Series on Semiconductor Parameters, vol. 2, p. 62. World Scientific, Singapore (1999)

29. Ko, Y., Sengupta, S., Tomasulo, S., Dutta, P., Wilke, I.: Emission of terahertz-frequency electromagnetic radiation from bulk Ga$_x$In$_{1-x}$As crystals. Phys. Rev. B **78**, 035201 (2008)

30. Baker, C., Gregory, I.S., Tribe, W.R., Bradley, I.V., Evans, M.J., Withers, M., Taday, P.F., Wallace, V.P., Linfield, E.H., Davies, A.G., Missous, M.: THz-pulse imaging with 1.06 μm laser excitation. Appl. Phys. Lett. **83**, 4113 (2003)

31. Suzuki, M., Tonouchi, M.: Fe-implanted InGaAs emitters for 1.56 mm wavelength excitation. Appl. Phys. Lett **86**, 1–051104 (2005)

Chapter 2
Generation of Sub-Picosecond Terahertz Pulses

In this chapter we will review the theoretical models of the various THz generation mechanisms from the surface of semiconductor materials. THz generation in semiconductors can proceed by the transport of photoexcited free carriers (current-surge effect) [1], or by difference frequency mixing (optical rectification) [2]. The current-surge effect is brought about by the acceleration of photocarriers in an electric field inside the semiconductor material. The field may be a built-in electric field inside the semiconductor, or may be externally applied (such as in the case of a photoconducting (PC) antenna). In the former case, the built-in electric field inside the semiconductor has two origins: (1) Surface depletion field caused by the Fermi level pinning at the semiconductor-air interface [3, 4], or (2) Photo-Dember field originating from the difference in diffusion constants of electrons and holes [3, 5]. In Sect. 2.1 we present a review of the past research done on $Ga_xIn_{1-x}As$ based THz emitters. Section 2.2 presents a theoretical model for the THz emission mechanism from an externally biased PC antenna for both cases of photoexcitation at normal and oblique incidence. In this section we have worked out in detail the mathematical framework of the theoretical model of THz emission from PC antennas that was developed earlier by several research groups [6–9]. The results obtained in this section have been used in Chap. 5 for modeling of the experimental data on THz emission from a $Ga_{0.69}In_{0.31}As$ based PC antenna. We conclude this chapter with Sect. 2.3 presenting a brief review of the theoretical models of the THz emission mechanism due to both current-surge and optical rectification effects.

2.1 Terahertz Emission from $Ga_xIn_{1-x}As$: A Review

As mentioned in Chap. 1, the III–V ternary alloy semiconductor $Ga_xIn_{1-x}As$ is an interesting choice for THz emitter materials driven by femtosecond fiber lasers with

S. Sengupta, *Characterization of Terahertz Emission from High Resistivity Fe-doped Bulk Ga$_{0.69}$In$_{0.31}$As Based Photoconducting Antennas*, Springer Theses, DOI: 10.1007/978-1-4419-8198-1_2, © Springer Science+Business Media, LLC 2011

emission wavelengths at 1.1 μm (photon energy 1.12 eV), since its band gap can be tuned from 0.36 to 1.42 eV by variation of Ga mole fraction from x = 0 to x = 1 [10]. However, since ternary semiconductors such as $Ga_xIn_{1-x}As$ are extremely difficult to grow and are not commercially available [11], very limited research has been carried out on the THz emission from this material. In the following sections we will review the past research conducted on THz emission from both $Ga_xIn_{1-x}As$ based photoconducting antenna emitter and surface emitters.

2.1.1 $Ga_xIn_{1-x}As$ Based Photoconducting Antenna Emitters

Almost all previous research on THz emission from $Ga_xIn_{1-x}As$ based PC antennas was focused on dipole antennas fabricated on $Ga_xIn_{1-x}As$ thin films grown by molecular beam epitaxy (MBE) on binary substrates. The primary challenge in fabricating large or semi-large aperture PC antenna based on $Ga_xIn_{1-x}As$ lies in obtaining high resistivity and good crystalline quality in bulk $Ga_xIn_{1-x}As$. Unlike low temperature grown (LT) GaAs, which can be grown in a semi-insulating (SI) form, LT $Ga_xIn_{1-x}As$ produces material of far lower resistivity or worse crystallinity [12]. Alternative approaches to increase the resistivity of $Ga_xIn_{1-x}As$ have been investigated by various research groups. Sukhotin et al. [13] reported high resistance (3.8×10^4 Ω) and high crystalline quality with picosecond carrier lifetimes in ErAs incorporated GaInAs with Be compensation. Baker et al. [14, 15] reported pulsed THz emission from MBE grown LT $Ga_{0.7}In_{0.3}As$ based PC antennas excited at 1.06 μm. In their work $Ga_{0.7}In_{0.3}As$ were grown on semi-insulating GaAs substrates at a nominal 230°C growth temperature and annealed ex situ following growth. Photoreflectance measurements indicated carrier lifetimes around 550 fs.

Ion irradiation is known as an alternative technique to increase resistivity and reduce carrier lifetimes in $Ga_xIn_{1-x}As$. Chimot et al. and Mangeney et al. [16, 17] demonstrated THz wave emission from heavy ion [Br^+ and Au^+] irradiated $Ga_{0.47}In_{0.53}As$ PC antennas triggered by 1.55 μm optical pulses. Heavy ion irradiation in $Ga_{0.47}In_{0.53}As$ resulted in a sheet resistance of $\sim 10^5$ Ω/square with carrier lifetimes less than 1 ps. In their work Chimot et al. [16] also demonstrated saturation of the emitted THz field with the increase in excitation pulse energy. The observed saturation was mainly due to the applied bias field screening by the excited photocarriers.

Suzuki et al. [18] reported THz emission from Fe ion implanted $Ga_{0.47}In_{0.53}As$ PC antennas triggered by 1.56 μm excitation pulses. The radiated field from the Fe ion implanted $Ga_{0.47}In_{0.53}As$ had a reduced pulse width compared to that obtained from the unimplanted emitters. This reduction in pulse width was attributed to the rapid relaxation of the carriers from the conduction state to the trapped state connected with the Fe ions and vacancies. It was also possible to apply a much higher bias voltage (~ 12 kV/cm) in the ion implanted material due to the increased resistivity. A linear dependence of the emitted THz radiation on the applied bias voltage was observed up to 12 kV/cm at an excitation power of 5 mW. As before, the THz emission was observed to saturate at higher pump powers. Suzuki et al. [19] also reported THz emission from unbiased $Ga_{0.47}In_{0.53}As$ and $Ga_{0.4}In_{0.6}As$ surfaces excited at 1.56 and 1.05 µm respectively. Similar saturation effects of the emitted THz field with increasing pump power were observed in this case also.

Takazato et al. [20] reported THz emission from PC antennas based on MBE grown LT $Ga_xIn_{1-x}As$ [x = 0.55, 0.47] layers on semi-insulating InP substrates. They demonstrated that reduction of In content in Be doped LT $Ga_xIn_{1-x}As$ results in increased resistivity of the material. In their experiments, resistivity as high as 760 Ω-cm was realized in $Ga_{0.55}In_{0.45}As$ based PC antennas. Such high resistivity made it possible to apply a bias of 60 kV/cm without device breakdown. It was observed that the peak amplitude of the radiated THz emission increased sub-linearly with excitation power and linearly with the bias voltage. LT $Ga_{0.55}In_{0.45}As$ was reported to be much superior to the LT $Ga_{0.47}In_{0.53}As$.

In all of the above cases reviewed, the THz emitters were based on MBE or MOCVD (Metal Organic Chemical Vapor Deposition) grown $Ga_xIn_{1-x}As$ layers on semi insulating substrates. While increase in resistivity was achieved by ion implantation and Be doping in LT $Ga_xIn_{1-x}As$, very high values such as those reported for LT GaAs (a common emitter material for large aperture dc-biased PC antenna) have not been reported to date. Further, the attempts to increase the resistivity of $Ga_xIn_{1-x}As$ by heavy-ion irradiation and ion implantation have only been partially successful since these processes result in low carrier mobility [16] that is detrimental to the THz emission. A comparison between the material properties of undoped $Ga_xIn_{1-x}As$, ion implanted and Be doped LT $Ga_xIn_{1-x}As$, SI GaAs, LT GaAs, and the high resistivity Fe-doped $Ga_{0.69}In_{0.31}As$ employed for the present experimental work, have been provided in Table 2.1. From Table 2.1, we see that compared to different types of PC antenna materials, the Fe-doped $Ga_{0.69}In_{0.31}As$ bulk crystal substrate employed in our research simultaneously exhibits the high resistivity, high mobility and sub-picosecond carrier lifetimes that are much desired for the superior performance of photoconducting antenna emitters.

2.1.2 $Ga_xIn_{1-x}As$ Surface Emitters

Ko et al. [24, 25] studied in detail the THz radiation emission from electrically unbiased bulk $Ga_xIn_{1-x}As$ crystals, with varying Ga mole fraction ($0 \leq x \leq 0.64$), photoexcited at 800 nm. Both optical rectification of the femtosecond NIR laser pulses, and current surge effects were found to be responsible for THz radiation emission from bulk $Ga_xIn_{1-x}As$ crystals. The THz emission due to transient photocurrents in $Ga_xIn_{1-x}As$ can be due to acceleration of photocarriers in a surface accumulation field or a Photo-Dember field depending on the Ga mole fraction x. In their calculations, Ko et al. showed that for $x < 0.2$ the Photo-Dember field dominates whereas for higher mole fractions the surface accumulation field takes over

Table 2.1 Comparison of properties between different types PC antenna materials

Material	Band gap (eV)	DC Hall mobility [cm^2/(Vs)]	Carrier lifetime (τ) [ps]	Max. bias voltage (E_b) [KV/cm]	Resistivity (sheet resistance)
LT-GaAs [21] T_g= 175–250°C	1.4	2250–3000	0.28–0.66	6	$>10^6$ Ω-cm
SI GaAs	1.4 [10]	$\geq$3000 [22]	10^5 [23]	–	$\geq$5 $\times$ 10^7 Ω-cm [22]
As-grown $Ga_{0.70}In_{0.30}As$ [24]	0.59	2058	0.2^a	–	19.5 Ω-cm
ErAs: $Ga_{0.47}In_{0.53}As$ [13]	0.75	–	0.8	–	100 Ω-cm
$Ga_{0.70}In_{0.30}As$ [14]	0.59	–	0.55 [14]	–	–
$Ga_{0.55}In_{0.45}As$ [20]	0.84	26 [20]	–	60 [20]	760 Ω-cm [20]
$Ga_{0.47}In_{0.53}As$ Dosage: 1 $\times$ 10^{12}/cm^2 Br^+ ions Energy 11 MeV [16]	0.75	490	<0.2	6.4	(0.3 $\times$ 10^5 Ω/square)
Fe implanted $Ga_{0.47}In_{0.53}As$ Dosage: 1 $\times$ 10^{15}/cm^2 Energy 340 keV [18]	0.75	1500	0.57	12.5	(6.13$\times$$10^3$ Ω/square)
Materials developed for this research SI $Ga_{0.69}In_{0.31}As$	1.0 [10]	2395	0.3^a	12.5	1.6 $\times$ 10^7 Ω-cm

[a] Carrier lifetime was measured using transient photo-reflectivity measurement described in Sect. 4.3 of Chap. 4

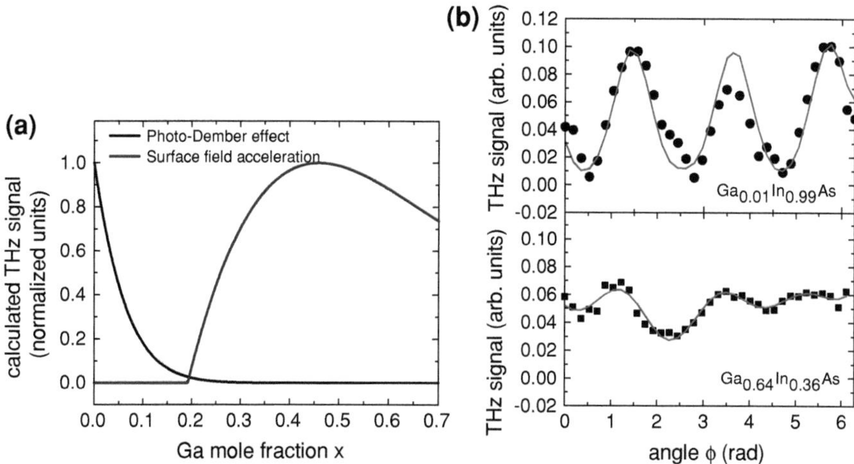

Fig. 2.1 **a** Calculations of the terahertz signal due to the photo-Dember effect and surface-field acceleration in Ga$_x$In$_{1-x}$As as a function of Ga mole fraction x. Calculated terahertz signals are normalized to 1 [24, 25]. **b** THz amplitude as a function of Azimuthal angle φ for Ga$_{0.01}$In$_{0.99}$As and Ga$_{0.64}$In$_{0.36}$As

(Fig. 2.1a). Further, the magnitude of THz emission was found to change periodically as a function of angle φ between the linear polarization of the excitation laser beam and the crystallographic orientation of the Ga$_x$In$_{1-x}$As surface as the samples were rotated about their surface normal (Fig. 2.1b).

Such periodic oscillations of the emitted THz signal can be attributed to the optical rectification of the incident femtosecond near-infrared laser pulses at the Ga$_x$In$_{1-x}$As surface, as explained in Sect. 2.3.2. The amplitude of oscillations was observed to be larger for Ga$_{0.01}$In$_{0.99}$As than Ga$_{0.64}$In$_{0.36}$As (Fig. 2.1b). The THz emission at the oscillation minima can be explained by transient photocurrents. The measurements for the entire compositional range by Ko et al. revealed that THz generation due to optical rectification decreases in Ga$_x$In$_{1-x}$As with increasing mole fraction x while THz generation due to transient photocurrents increases with increasing Ga mole fraction. The researchers attributed this experimental observation to the reduced efficacy of the THz optical rectification process due to decreasing microcrystal size in the polycrystalline Ga$_x$In$_{1-x}$As samples with the increase in Ga mole fraction. The overall terahertz emission from bulk Ga$_x$In$_{1-x}$As crystals was found to be maximized in the $x \approx$ 0.1–0.3 compositional range.

2.2 Theoretical Model of THz Emission from Semi-Large Aperture DC Biased Photoconducting Antenna Emitters

A large aperture PC antenna comprises of a planar semiconductor substrate (with dielectric constant ϵ) onto which parallel conducting electrodes are adhered.

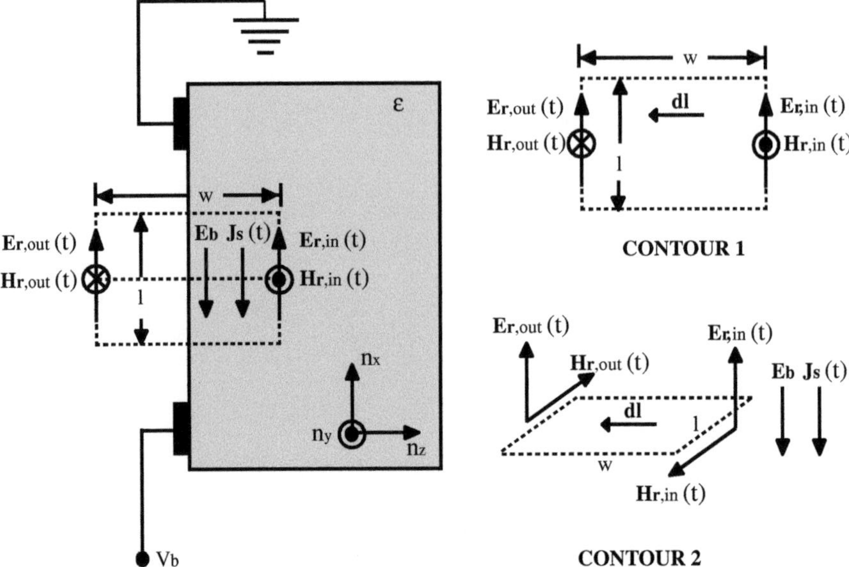

Fig. 2.2 Schematic diagram of semi-large aperture photoconducting antenna excited at normal incidence and biased at a voltage V_b [6]. The diagram depicts the relevant fields and currents: the bias electric field $\mathbf{E}_b$; the time dependent inward and outward radiating electric and magnetic fields $\mathbf{E}_{r,\text{in}}(t), \mathbf{E}_{r,\text{out}}(t), \mathbf{H}_{r,\text{in}}(t)$, and $\mathbf{H}_{r,\text{out}}(t)$ respectively and the time varying surface current $\mathbf{J}_s(t)$. Two orthogonal surfaces, l units long and w units wide, with corresponding contours partially inside the semiconductor are also shown

We recall from Chap. 1, that large aperture in this context means that the transverse dimensions (or more specifically, the electrode spacing) of the antenna are large (typically greater than 0.3 mm) relative to the center wavelength of the emitted THz radiation (1 THz = 300 μm). The term "semi-large" has been used in literature [9, 26] to loosely categorize PC antennas with electrode gap spacing between 0.1 and 1 mm. They offer a good compromise between microscopically small (<100 μm) and millimeter (>1 mm) sized structures in that they are relatively easy to fabricate and their spectral characteristics are less dependent on antenna geometry (unlike micro-dipole antennas), while at the same time, high THz powers can be produced by using unamplified mode-locked femtosecond lasers (unlike large gap sources). From the theoretical standpoint, we can still apply the theoretical model of THz emission from large aperture PC antennas [6–8] to the semi-large aperture PC antennas without any loss of generalization as we will see in the following paragraphs.

Figure 2.2 shows a PC antenna with an applied bias voltage $\mathbf{V_b}$ (and corresponding bias field $\mathbf{E_b}$) photoexcited at normal incidence. Figure 2.2 also depicts the relevant time dependent quantities such as: the inward and outward radiated electric fields, $\mathbf{E_{r,in}}(t)$ and $\mathbf{E_{r,out}}(t)$, the inward and outward magnetic fields $\mathbf{H_{r,in,}}(t)$ and $\mathbf{H_{r,out}}(t)$ (pointing in and out of the page respectively), and the surface current $\mathbf{J}_s(t)$. The radiated fields are assumed to be plane waves propagating away from the conducting surface. Two mutually orthogonal surfaces, l units long and w units wide, with corresponding contours partially inside the semiconductor surface are also depicted. Three unit normal vectors $\mathbf{n}_x$, $\mathbf{n}_y$, and $\mathbf{n}_z$ that are mutually orthogonal are shown in Fig. 2.2.

Using Maxwell's equations we next proceed to derive the relationships among all these quantities. In spite of the time dependent nature of the physical quantities depicted in Fig. 2.2, it can be shown that the boundary conditions on the electric and magnetic fields reduce to those of the steady state. The derivations in the remainder of this section follow the theoretical model outlined in [6–9].

The boundary condition on the electric fields can be obtained from Faraday's law [27],

$$\nabla \times \mathbf{E} = -\frac{\partial \mathbf{B}}{\partial t} = -\frac{\mu_0 \mu_r \partial \mathbf{H}}{\partial t} \tag{2.1}$$

where μ_0 and μ_r are the magnetic permeability of free space and the of the PC antenna substrate material respectively. Since the material is non-magnetic, we assume $\mu_r = 1$. Using Stoke's theorem [28], we can integrate Eq. 2.1 over a flat region perpendicular to the PC antenna surface with unit normal $\mathbf{n}_y$.

$$\int (\nabla \times \mathbf{E}) \cdot d\mathbf{a} = \oint \mathbf{E} \cdot d\mathbf{l} = -\mu_0 \frac{\partial}{\partial t} \int \mathbf{H} \cdot d\mathbf{a} = -\mu_0 \frac{\partial}{\partial t} \int \mathbf{H} \cdot \mathbf{n}_y da. \tag{2.2}$$

The contour integral in Eq. 2.2 is performed along the perimeter of the region depicted as CONTOUR 1 in Fig. 2.2. Since the distance l is taken to be larger than the center wavelength of the emitted radiation, but smaller than the antenna gap spacing, the electric fields are assumed to be uniform over the respective sides of length l along the contour of integration. Thus Eq. 2.2 reduces to

$$\left(E_{r,\text{in}}(t) - E_{r,\text{out}}(t)\right)l = -lw\mu_0 \frac{\partial}{\partial t}\left(H_{r,\text{in}}(t) - H_{r,\text{out}}(t)\right)$$
$$\left(E_{r,\text{in}}(t) - E_{r,\text{out}}(t)\right) = -w\mu_0 \frac{\partial}{\partial t}\left(H_{r,\text{in}}(t) - H_{r,\text{out}}(t)\right). \tag{2.3}$$

Since we assume that the right hand side of Eq. 2.3 is bounded, in the limit of width w becoming very small, Eq. 2.3 reduces to

$$\lim_{w \to 0}\left(E_{r,\text{in}}(t) - E_{r,\text{out}}(t)\right) = 0. \tag{2.4}$$

The radiated electric fields are thus equal at the photoconductor surface

$$\mathbf{E}_{r,\text{in}}(t) = \mathbf{E}_{r,\text{out}}(t). \tag{2.5}$$

Equation 2.5 further reduces to the steady state boundary condition [29]

$$\mathbf{n}_z \times \left(\mathbf{E}_{r,\text{in}}(t) - \mathbf{E}_{r,\text{out}}(t)\right) = 0. \tag{2.6}$$

By a similar analogy, the boundary condition on the radiated magnetic fields can be derived from the Ampere-Maxwell law [27]

$$\nabla \times \mathbf{H} = \left(\mathbf{J} + \frac{\partial \mathbf{D}}{\partial t} \right) \tag{2.7}$$

where $\mathbf{J}$ is the current density and $\mathbf{D}$ is the electric displacement (i.e. $D = \epsilon E = \epsilon_0 \epsilon_r E$). Here ϵ and ϵ_r are the dielectric permittivity and the relative dielectric permittivity (or the dielectric constant) of the PC antenna substrate respectively and ϵ_0 is the permittivity of free space. Applying Stoke's theorem to Eq. 2.7 we obtain

$$\oint \mathbf{H} \cdot d\mathbf{l} = \int \left(\mathbf{J} + \frac{\partial \mathbf{D}}{\partial t} \right) \cdot \mathbf{n}_x da \tag{2.8}$$

where the contour integral is performed along the region depicted as CONTOUR 2 in Fig. 2.2.

$$-\left(H_{r,\text{out}}(t) + H_{r,\text{in}}(t) \right)l = -l \int_0^w J(z,t)dz + lw\frac{\partial}{\partial t}\left(E_{r,\text{out}}(t) + \epsilon E_{r,\text{in}}(t) \right)$$

$$\left(H_{r,\text{out}}(t) + H_{r,\text{in}}(t) \right) = \int_0^w J(z,t)dz - w(1+\epsilon)\frac{\partial}{\partial t}E_{r,\text{in}}(t) \tag{2.9}$$

$$\left(H_{r,\text{out}}(t) + H_{r,\text{in}}(t) \right) = J_s(t) - w(1+\epsilon)\frac{\partial}{\partial t}E_{r,\text{in}}(t).$$

Here $\mathbf{J}_s(t)$ is the surface current density which is related to the volume current density $\mathbf{J}(t)$ by

$$J_s(t) = \int_0^\delta J(z,t)dz \tag{2.10}$$

where δ is the distance inside the photoconductor that excited carriers exist (or in other words the penetration depth of the laser beam) and dz is the incremental distance into the surface along the unit normal $\mathbf{n_z}$. It is to be noted that $\mathbf{J_s}$ is a current per unit length (Am^{-1}) i.e. a linear current density as opposed to a bulk current density (Am^{-2}) occurring in Ohm's law. Since $(\partial/\partial t)E_{r,\text{in}}(t)$ is bounded, the second term on the right hand side of Eq. 2.9 is negligible in the limit of small w. Thus Eq. 2.10 becomes

$$\lim_{w \to 0} \left(H_{r,\text{in}}(t) + H_{r,\text{out}}(t) \right) = J_s(t). \tag{2.11}$$

Equation 2.11 can be expressed as the steady state boundary condition on magnetic fields as:

$$\left(\mathbf{H}_{r,\text{out}}(t) - \mathbf{H}_{r,\text{in}}(t) \right) \times \mathbf{n_z} = \mathbf{J_s}(t) \tag{2.12}$$

The radiated electric and magnetic fields are further related by the dielectric constant of the PC antenna substrate ϵ_r and free space resistance Z_0 as [30]

$$\mathbf{H}_{r,\mathrm{in}}(t) \times \mathbf{n_z} = \frac{\sqrt{\epsilon_r}}{Z_0} \mathbf{E}_{r,\mathrm{in}}(t) \tag{2.13}$$

and

$$\mathbf{H}_{r,\mathrm{out}}(t) \times \mathbf{n_z} = \frac{1}{Z_0} \mathbf{E}_{r,\mathrm{out}}(t). \tag{2.14}$$

Equations 2.6, 2.12, 2.13 and 2.14 are also valid for the general case of oblique incident of the optical pulse [31]. Using these equations we proceed to derive the functional form of the transmitted and the reflected THz electric field for the cases of normal and oblique incidence of the optical beam respectively.

2.2.1 Case I: Normal Incidence

In this section we proceed to derive the functional form of the THz electric field transmitted into the photoconductor for the configuration presented in Fig. 2.2. From Eqs. 2.5, 2.12, 2.13, and 2.14, the surface current density $\mathbf{J_s}(t)$ can be expressed as a function of the inward radiated electric field $\mathbf{E}_{r,\mathrm{in}}(t)$ as

$$\mathbf{J_s}(t) = -\frac{\left(1 + \sqrt{\epsilon_r}\right)}{Z_0} \mathbf{E}_{r,\mathrm{in}}(t) \tag{2.15}$$

Thus the radiated electric field inside the semiconductor has similar time dependence as the surface current density $\mathbf{J_s}(t)$. This characteristic was previously observed in the radiated electric field from PC antennas at observation points in or near the near-field region [32–34]. From Ohm's law, the surface current density $\mathbf{J_s}(t)$ is given by:

$$\mathbf{J_s}(t) = \sigma_s(t)\left(\mathbf{E_b} + \mathbf{E}_{r,\mathrm{in}}(t)\right) \tag{2.16}$$

where $\sigma_s(t)$ is the surface conductivity of the emitting antenna and is defined as [6, 33]:

$$\sigma_s(t) = \frac{e(1 - R)}{\hbar\omega} \int\limits_{-\infty}^{t} dt'\, \mu(t - t') I_{\mathrm{opt}}(t') \exp\frac{(-t - t')}{\tau_c}. \tag{2.17}$$

Here, e is the electron charge, R is the optical reflectivity of illuminated surface of the photoconductor, $I_{\mathrm{opt}}(t')$ and $\mu(t - t')$ are the optical pulse intensity and carrier mobility as functions of time, $\hbar\omega$ is the photon energy of the laser pulse, and τ_c is the lifetime of the excited carriers. For the present derivation we assume carrier mobility μ to be constant. Further, μ is taken to be the electron mobility since the hole mobility is much lower and can be neglected [35]. Equation 2.17 can be simplified as

$$\sigma_s(t) = \frac{e(1-R)\mu I_0}{\hbar\omega} \int_{-\infty}^{t} dt' g(t') \exp\left(-\frac{t-t'}{\tau_c}\right) = e\mu n(t) \tag{2.17a}$$

where I_0 and $g(t')$ are the intensity and temporal profile of the laser pulse respectively and $n(t)$ is the time dependent photocarrier density given by

$$n(t) = \frac{(1-R)I_0}{\hbar\omega} \int_{-\infty}^{t} dt' g(t') \exp\left(-\frac{t-t'}{\tau_c}\right) = G_0 \int_{-\infty}^{t} dt' g(t') \exp\left(-\frac{t-t'}{\tau_c}\right)$$

$$\tag{2.18}$$

where $G_0 = (1-R)(I_0/\hbar\omega)$ is the absorbed photon density per unit area, per unit time, and $I_0 = \mathbf{P}_{\mathrm{opt}}/A_{\mathrm{opt}}$ is the intensity of the optical light. $\mathbf{P}_{\mathrm{opt}}$, and A_{opt} are the incident laser power and the area of illumination respectively. From Eqs. 2.15 and 2.16, the surface current $\mathbf{J_s}(t)$, and the inward radiated field $\mathbf{E}_{r,\mathrm{in}}(t)$ at the emitter can be expressed as [6]

$$\mathbf{J_s}(t) = \mathbf{E_b}\frac{\left(1+\sqrt{\epsilon_r}\right)\sigma_s(t)}{\sigma_s(t)Z_0 + \left(1+\sqrt{\epsilon_r}\right)} \tag{2.19a}$$

$$\mathbf{E}_{r,\mathrm{in}}(t) = -\mathbf{E_b}\frac{\sigma_s(t)Z_0}{\sigma_s(t)Z_0 + \left(1+\sqrt{\epsilon_r}\right)}. \tag{2.19b}$$

Using Eqs. 2.1 through Eqs. 2.19a and 2.19b we now proceed to calculate the peak value of the THz electric field and the average radiated THz power in the far field

2.2.2 Case II: Oblique Incidence

In this section we derive the functional form of the radiated THz electric field in the direction of specular reflection in the near-field region of the photoconductor for the case of oblique incidence of the optical pulse (Fig. 2.3). As before, we assume plane wave-fronts propagating away from the photoconductor surface. Here θ_{opt} is the angle of incidence of the optical pulse. The time-varying surface current in the photoconductor emits THz radiation in the forward and backward directions at angles θ_2 and θ_1 to the surface normal respectively. The relationship between the angle of incidence and the angle of reflection and refraction can be expressed by generalized Snell's law equation as:

$$n_1\left(\omega_{\mathrm{opt}}\right)\sin\theta_{\mathrm{opt}} = n_1(\omega_{\mathrm{THz}})\sin\theta_1 = n_2(\omega_{\mathrm{THz}})\sin\theta_2 \tag{2.20}$$

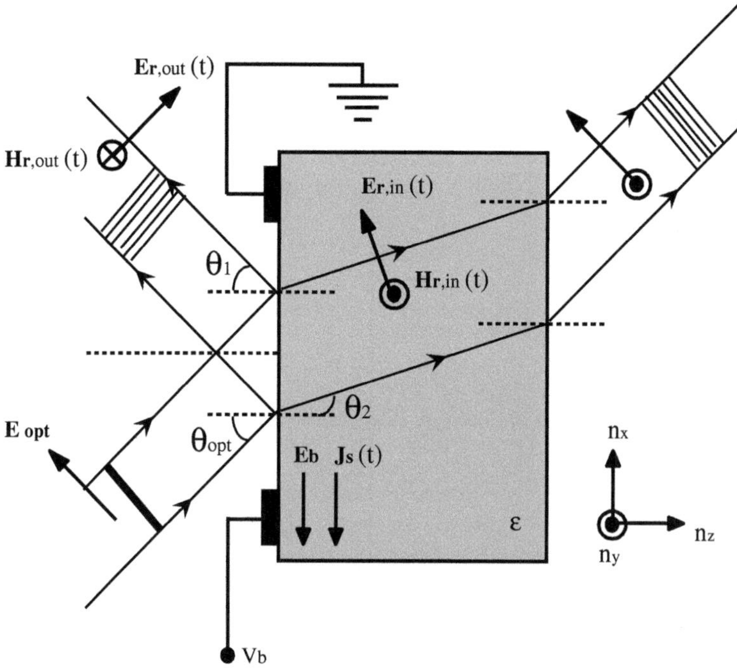

Fig. 2.3 Schematic diagram of semi-large aperture photoconducting antenna excited at oblique incidence and biased at a voltage V_b

where $n_1(\omega_{opt})$ and $n_1(\omega_{THz})$ are indices of refraction of the medium outside the photoconductor at optical and THz frequencies respectively and $n_2(\omega_{THz})$ is the index of refraction of material of the photoconductor at the THz frequency. For the case where the medium outside the photoconductor is air, $n_1(\omega_{opt}) = n_1(\omega_{THz}) = 1$ and $\theta_1 = \theta_{opt}$ or in other words the backward THz pulse is emitted in the direction of the specular reflection of the optical pulse. From Eq. 2.6 we have

$$\mathbf{E}_{r,out}(t) \cos\theta_1 = \mathbf{E}_{r,in}(t) \cos\theta_2 \tag{2.21}$$

$$\mathbf{E}_{r,in}(t) = \mathbf{E}_{r,out}(t)\frac{\cos\theta_1}{\cos\theta_2}. \tag{2.22}$$

Using boundary condition (2.12) we have

$$\mathbf{H}_{r,out}(t) + \mathbf{H}_{r,in}(t) = -\mathbf{J_s}(t). \tag{2.23}$$

Further, Eqs. 2.13 and 2.14 yield

$$\mathbf{H}_{r,in}(t) = \frac{\sqrt{\epsilon_r}}{Z_0}\mathbf{E}_{r,in}(t) \tag{2.24}$$

and

$$\mathbf{H}_{r,\text{out}}(t) = \frac{1}{Z_0}\mathbf{E}_{r,\text{out}}(t).$$ (2.25)

From Eqs. 2.22, 2.23, 2.24 and 2.25 $\mathbf{J_s}(t)$ can be expressed as a function of the backward radiated electric field $\mathbf{E}_{r,\text{out}}(t)$d, and angles θ_1, and θ_2 as

$$\mathbf{J_s}(t) = -\left(\frac{\sqrt{\epsilon_r}}{Z_0}\frac{\cos\theta_1}{\cos\theta_2} + \frac{1}{Z_0}\right)\mathbf{E}_{r,\text{out}}(t).$$ (2.26)

From Ohm's law, the surface current density $\mathbf{J}_s(t)$ is given by

$$\mathbf{J_s}(t) = \sigma_s(t)\left(\mathbf{E_b} + \mathbf{E}_{r,\text{in}}(t)\cos\theta_2\right)$$ (2.27)

where $\sigma_s(t)$ is the surface conductivity of the emitting antenna and is defined as [31]:

$$\sigma_s(t) = \frac{e(1-R)\mu\cos\theta_1}{\hbar\omega}\int_{-\infty}^{t} dt' I_{\text{opt}}(t')\exp\left(-\frac{t-t'}{\tau_c}\right).$$ (2.28)

Using Eqs. 2.21, 2.26 and 2.27, the surface current $\mathbf{J_s}(t)$ and the backward radiated THz field $\mathbf{E}_{r,\text{out}}(t)$ can be expressed as functions of the applied bias field $\mathbf{E_b}$, surface conductivity $\sigma_s(t)$, and angles θ_1, and θ_2 as [31].

$$\mathbf{J_s}(t) = \mathbf{E_b}\frac{\left(1 + \sqrt{\epsilon_r}\frac{\cos\theta_1}{\cos\theta_2}\right)\sigma_s(t)}{\sigma_s(t)Z_0\cos\theta_1 + \left(1 + \sqrt{\epsilon_r}\frac{\cos\theta_1}{\cos\theta_2}\right)}$$ (2.29a)

$$\mathbf{E}_{r,\text{out}}(t) = -\frac{\mathbf{E_b}}{\cos\theta_1}\frac{\sigma_s(t)Z_0}{\left(\sigma_s(t)Z_0 + \frac{1}{\cos\theta_1} + \frac{\sqrt{\epsilon_r}}{\cos\theta_2}\right)}.$$ (2.29b)

It is to be noted that for normal incidence of the optical pulse (i.e. $\theta_{\text{opt}} = \theta_1 = \theta_2 = 0$ and $\cos\theta_1 = \cos\theta_2 = 1$), Eqs. 2.26, 2.29a and 2.29b reduce to the form of Eqs. 2.15, 2.19a and 2.19b respectively.

2.2.3 Calculation of Peak THz Electric Field

2.2.3.1 Case I: Normal Incidence

From Eqs. 2.19a and 2.19b, the maximum surface current and the peak radiated field inside the antenna can be expressed as [6]:

$$\mathbf{J_{s,max}} = \mathbf{E_b}\frac{\left(1 + \sqrt{\epsilon_r}\right)\sigma_{s,\text{max}}}{\sigma_{s,\text{max}}Z_0 + \left(1 + \sqrt{\epsilon_r}\right)}$$ (2.30a)

$$\mathbf{E}_{\mathbf{r},\mathrm{in,max}} = -\mathbf{E}_{\mathbf{b}}\frac{\sigma_{s,\mathrm{max}}Z_0}{\sigma_{s,\mathrm{max}}Z_0 + \left(1 + \sqrt{\epsilon_r}\right)} \qquad (2.30\mathrm{b})$$

where $\sigma_{s,\mathrm{max}}$ is the peak surface conductivity expressed as [6]:

$$\sigma_{s,\mathrm{max}} = \frac{e(1 - R)\mu F_{\mathrm{opt}}}{\hbar\omega} \qquad (2.31)$$

Here, F_{opt} is the optical excitation fluence. Equations 2.30b and 2.31 predict that as the optical fluence (and thus, the surface photoconductivity) becomes large, the inward radiated electric field $\mathbf{E}_{\mathbf{r},\mathrm{in,max}}$ approaches the bias field $\mathbf{E}_{\mathbf{b}}$ in magnitude. From Maxwell's equation in the Coulomb gauge, the radiated field $\mathbf{E}_{\mathbf{r}}(\mathbf{r},t)$ as a function of time t at a displacement $\mathbf{r}$ from the antenna center is given by [36]:

$$\mathbf{E}_{\mathbf{r}}(\mathbf{r},t) = -\frac{1}{4\pi\epsilon_0 c^2}\frac{\partial}{\partial t}\int \frac{\mathbf{J}_{\mathbf{s}}(\mathbf{r},t - (|\mathbf{r} - \mathbf{r'}|/c))}{|\mathbf{r} - \mathbf{r'}|}\,da \qquad (2.32)$$

In the far field geometry, $|\mathbf{r'}| \ll |\mathbf{r}|$; thus

$$|\mathbf{r} - \mathbf{r'}| = r\left(1 - \frac{\mathbf{n}\cdot\mathbf{r'}}{r}\right) \approx r. \qquad (2.33)$$

Also, since the gap between the electrodes of the PC antenna is assumed to be uniformly illuminated by the optical pulse, the surface current $\mathbf{J}_{\mathbf{s}}$ can be assumed to spatially constant at all points on the surface of the emitter. The radiated electric field can be written as:

$$\mathbf{E}_{\mathbf{r}}(\mathbf{r},t) = -\frac{1}{4\pi\epsilon_0 c^2}\frac{A}{(x^2 + y^2 + z^2)^{1/2}}\frac{\partial}{\partial t}\mathbf{J}_{\mathbf{s}}\left(t - \frac{r}{c}\right) \qquad (2.34)$$

where A is the electrode gap spacing of the PC antenna. For radiation emitted and detected on z axis (i.e., $x = y = 0$) and $t \rightarrow t - z/c$ Eq. 2.34 reduces to

$$\mathbf{E}_{\mathbf{r}}(\mathbf{r},t) \cong -\frac{1}{4\pi\epsilon_0 c^2}\frac{A}{z}\frac{\partial}{\partial t}\mathbf{J}_{\mathbf{s}}(t). \qquad (2.35)$$

Further, since the surface current density increases to its maximum value on the order of the pulse-width of the radiated field τ_p, it can be assumed that

$$\left(\frac{\partial\mathbf{J}_{\mathbf{s}}}{\partial t}\right)_{\mathrm{max}} \cong \frac{\mathbf{J}_{\mathbf{s},\mathrm{max}}}{\tau_p}. \qquad (2.36)$$

Using Eqs. 2.33–2.36, Eq. 2.32 can be simplified to express the maximum value of the far-field radiated THz field as a function of the displacement along the center of the radiated beam z [6]

$$\mathbf{E}_{\text{r,far,max}}(z) \cong -\frac{1}{4\pi\epsilon_0 c^2}\frac{J_{\text{s,max}}}{\tau_p}\frac{A}{z} = \left|\mathbf{E_b}\frac{\sigma_{s,\text{max}}}{\sigma_{s,\text{max}}Z_0 + \left(1 + \sqrt{\epsilon_r}\right)}\cdot\frac{\left(1 + \sqrt{\epsilon_r}\right)A}{4\pi\epsilon_0 c^2\tau_p z}\right| \quad (2.37)$$

From Eq. 2.37 two important points should be noted:

- Since the peak surface conductivity $\sigma_{s,\text{max}}$ has a linear dependence on the incident optical fluence F_{opt} (for a given emitter material), then for a given bias voltage, the emitted THz radiation saturates for very high values of optical fluence
- At a low optical fluence, the radiated THz field varies linearly as the applied bias field $\mathbf{E_b}$ at least in the area of Ohm's law validity. At very high bias voltages, the field dependence of the conductivity σ_s has to be taken into account.

2.2.3.2 Case II: Oblique Incidence

Using Eq. 2.29a, the peak surface current density $\mathbf{J}_{\text{s,max}}$ can be expressed as

$$\mathbf{J}_{\text{s,max}} = \mathbf{E_b}\frac{\left(1 + \sqrt{\epsilon_r}\frac{\cos\theta_1}{\cos\theta_2}\right)\sigma_{s,\text{max}}}{\sigma_{s,\text{max}}Z_0\cos\theta_1 + \left(1 + \sqrt{\epsilon_r}\frac{\cos\theta_1}{\cos\theta_2}\right)} \quad (2.38)$$

where $\sigma_{s,\text{max}}$ is the peak surface conductivity given by

$$\sigma_{s,\text{max}} = \frac{e(1 - R)\mu\cos\theta_1 F_{\text{opt}}}{\hbar\omega}. \quad (2.39)$$

Using Eqs. 2.35, 2.36 and 2.38, the peak value of the radiated electric field can be expressed a function of the applied bias field $\mathbf{E_b}$, peak surface conductivity $\sigma_{s,\text{max}}$, electrode gap spacing of the photoconductor A, and angles θ_1, and θ_2 as

$$\mathbf{E}_{\text{r,far,max}}(z) \cong -\frac{1}{4\pi\epsilon_0 c^2}\frac{J_{\text{s,max}}}{\tau_p}\frac{A}{z} = \left|\mathbf{E_b}\frac{\left(1 + \sqrt{\epsilon_r}\frac{\cos\theta_1}{\cos\theta_2}\right)\sigma_{s,\text{max}}}{\sigma_{s,\text{max}}Z_0\cos\theta_1 + \left(1 + \sqrt{\epsilon_r}\frac{\cos\theta_1}{\cos\theta_2}\right)}\cdot\frac{A}{4\pi\epsilon_0 c^2\tau_p z}\right|$$

$$(2.40)$$

2.2.4 Calculation of Average Radiated THz Power

Since $\sigma_s(t) \propto F_{\text{opt}}$ as evident from Eqs. 2.28 and 2.31, in the low optical fluence regime (such as in our experiments) Eq. 2.19a reduces to:

$$\mathbf{J_s}(t) \cong \sigma_s(t)\mathbf{E_b} = e\mu_n(t)\mathbf{E_b} \quad (2.41)$$

where $n(t)$ is given by Eq. 2.18. Although the pulse shape of an ideal mode-locked laser has a Gaussian profile, a useful approximation can be obtained by assuming a square pulse shape such that [9]:

$$g(t) = \begin{cases} 1, & 0 < t < T_{\text{pulse}} \\ 0, & \text{elsewhere} \end{cases} \tag{2.42}$$

where, T_{pulse} is the pulse duration of the laser pulse. From Eq. 2.18, the value of photocarrier density is then given by [9]:

$$n(t) = \begin{cases} G_0 \tau_c \left(1 - e^{-\frac{t}{\tau_c}}\right), & 0 < t < T_{\text{pulse}} \\ G_0 \tau_c e^{-\frac{t}{\tau_c}}, & t > T_{\text{pulse}} \end{cases} \tag{2.43}$$

$$\frac{\partial n(t)}{\partial t} = \begin{cases} G_0 e^{-\frac{t}{\tau_c}}, & 0 < t < T_{\text{pulse}} \\ -G_0 e^{-\frac{t}{\tau_c}}, & t > T_{\text{pulse}} \end{cases} \tag{2.44}$$

Taking the time average of $\frac{\partial n(t)}{\partial t}$ over one period of the laser pulse we obtain

$$\left\langle \left(\frac{\partial n(t)}{\partial t}\right)^2 \right\rangle = \frac{1}{T_{\text{rep}}} \int_0^{T_{\text{rep}}} \left(\frac{\partial n(t)}{\partial t}\right)^2 dt = \frac{G_0^2}{T_{\text{rep}}} \frac{\tau_c}{2} \tag{2.45}$$

where, T_{rep} is the time period between successive laser pulses. We also assume that T_{rep} is much longer than the carrier lifetime τ_c and the laser pulse duration T_{pulse}. From Eq. 2.35, the THz emission in the far field is given by

$$\mathbf{E_r}(\mathbf{r}, t) \cong -\frac{1}{4\pi\epsilon_0 c^2} \frac{A}{z} e\mu \mathbf{E_b} \frac{\partial}{\partial t} n(t). \tag{2.46}$$

From Eqs. 2.45 and 2.46 and using Poynting's theorem [37], the average radiated THz power per unit area can be computed as

$$\langle \mathbf{P} \rangle = \frac{c\epsilon_0}{2} \langle \mathbf{E_r^2} \rangle \cong \left(\frac{c\epsilon_0}{2}\right) \left(\frac{e\mu \mathbf{E_b}}{4\pi\epsilon_0 c^2} \frac{A}{z}\right)^2 \left\langle \left(\frac{\partial n(t)}{\partial t}\right)^2 \right\rangle = \frac{e^2 \mu^2 \mathbf{E_b}^2 A^2}{32\pi^2 z^2 \epsilon_0 c^3} \left(\frac{G_0^2 \tau_c}{2T_{\text{rep}}}\right). \tag{2.47}$$

The average THz power per unit area is thus proportional to the square of both the applied field and laser power. It is also proportional to the carrier lifetime. However a long carrier lifetime reduces peak THz field (and thus the peak THz power) as is evident from Eq. 2.40 and, therefore, is detrimental to THz generation.

2.2.5 *Power Scaling of Terahertz Radiation with Antenna Gap Spacing*

In this section we proceed to compute the dependence of the THz power on the antenna gap spacing. The average THz power can be computed from Eqs. 2.47 and 2.18 as

$$
\begin{aligned}
\mathbf{P}_{\text{avg}} = \langle \mathbf{P} \rangle A &= \left(\frac{e^2 \mu^2 \mathbf{E}_b{}^2 A^2}{32\pi^2 z^2 \epsilon_0 c^3} \right) \left(\frac{G_0^2 \tau_c}{2 T_{\text{rep}}} \right) A' \\
&= \left(\frac{e^2 \mu^2 \mathbf{E}_b{}^2}{32\pi^2 z^2 \epsilon_0 c^3} \right) \left(\frac{(1-R)\mathbf{P}_{\text{opt}}}{\hbar\omega} \right)^2 \left(\frac{A}{A_{\text{opt}}} \right)^2 \left(\frac{\tau_c}{2 T_{\text{rep}}} \right) A'
\end{aligned}
\tag{2.48}
$$

where, A' is the beam diameter of the THz radiation at the detector. If we neglect the effects of diffraction, it can be assumed that $A' = A$, the area between the electrode gap spacing of the photoconductor.[1] Indeed, for large aperture (>1 mm) photoconductive antennas it has been shown before [6] that the higher frequency components of the emitted THz radiation diffract less, and are closer to the center of the radiated beam as the antenna gap spacing is increased. Also for our experimental conditions, $A \approx A_{\text{opt}}$, the area of illumination. Equation 2.48 can thus be simplified to

$$
\mathbf{P}_{\text{avg}} \approx \left(\frac{e^2 \mu^2 \mathbf{E}_b{}^2}{32\pi^2 z^2 \epsilon_0 c^3} \right) \left(\frac{(1-R)\mathbf{P}_{\text{opt}}}{\hbar\omega} \right)^2 \left(\frac{\tau_c}{2 T_{\text{rep}}} \right) A.
\tag{2.49}
$$

Thus, for a given photoconductor, if the applied bias field $\mathbf{E}_b$ and the incident laser power $\mathbf{P}_{\text{opt}}$ are kept constant, the average radiated THz power scales linearly as the area A between the electrode gap spacing of the PC antenna. In reality, however, the peak radiated THz electric field (and thus the peak power) does not scale linearly with the antenna gap spacing A as is evident from Eqs. 2.30b, 2.31, and 2.37.

$$
\mathbf{P}_{r,\text{in,max}} \propto \left| E_{r,\text{in,max}} \right|^2 A = A \left(\mathbf{E}_b \frac{\sigma_{s,\text{max}} Z_0}{\sigma_{s,\text{max}} Z_0 + \left(1 + \sqrt{\epsilon_r} \right)} \right)^2 = \mathbf{E}_b{}^2 A \left(\frac{1}{1 + \frac{\left(1 + \sqrt{\epsilon_r} \right)}{\sigma_{s,\text{max}} Z_0}} \right)^2.
\tag{2.50}
$$

Substituting the value of $\sigma_{s,\text{max}}$ from Eq. 2.31 in Eq. 2.50 and defining $F_{\text{opt}} = E_{\text{opt}}/A$ where E_{opt} is the pulse energy of the incident optical pulse, we obtain

---

[1] More precisely A is defined as a circular area having a diameter equal to the electrode gap spacing of the PC antenna.

$$P_{\text{r, in, max}} \propto \mathbf{E_b}^2 \frac{A}{\left(1 + \frac{\left(1+\sqrt{\epsilon_r}\right)}{\frac{e(1-R)\mu E_{\text{opt}}Z_0}{\hbar\omega}}A\right)^2} = \mathbf{E_b}^2 \frac{A}{(1+DA)^2} \qquad (2.51)$$

where

$$D = \frac{\left(1+\sqrt{\epsilon_r}\right)}{\frac{e(1-R)\mu E_{\text{opt}}Z_0}{\hbar\omega}}. \qquad (2.52)$$

By taking the derivative of Eq. 2.51 with respect to area, and equating it to zero, the value of the antenna electrode gap spacing that results in the maximum radiated pulse energy can be estimated. For a particular photoconductor and for constant values of bias electric field $\mathbf{E_b}$ and optical pulse energy E_{opt}, this particular area $A_{\max}$ is [7]

$$A_{\max} = \frac{1}{D} = \frac{e(1-R)\mu E_{\text{opt}}Z_0}{\hbar\omega\left(1+\sqrt{\epsilon_r}\right)}. \qquad (2.53)$$

As physically intuitive, this optimum area increases with increasing pulse energy.

Equations 2.1 through 2.53 describe the theoretical model of THz emission from voltage biased semi-large aperture PC antennas. We shall make use of the results derived here in Chap. 5 again for modeling the THz emission from a $Ga_{0.69}In_{0.31}As$ based PC antenna.

2.3 Semiconductor Surface Emitters

In Sect. 2.2, we presented a detailed mathematical framework for the theoretical model of the THz emission from semi-large aperture DC biased photoconducting antenna emitters. In this section we will review in brief the theory of THz emission from semiconductor surface emitters. Since the mathematical details have been presented earlier in the scientific literature [3, 25] we will only review the relevant results here.

2.3.1 Current-Surge Model

In an unbiased semiconductor surface emitter, THz emission proceeds by acceleration of photocarriers generated by the pump laser pulse in the built-in field inside the semiconductor. The built-in field inside the semiconductor is thought to have two origins: (1) the surface depletion field caused by the Fermi level pinning and (2) the Photo-Dember field originating from the difference in diffusion constants of electrons and holes [3]. A schematic diagram of both the processes is depicted in Fig. 2.4.

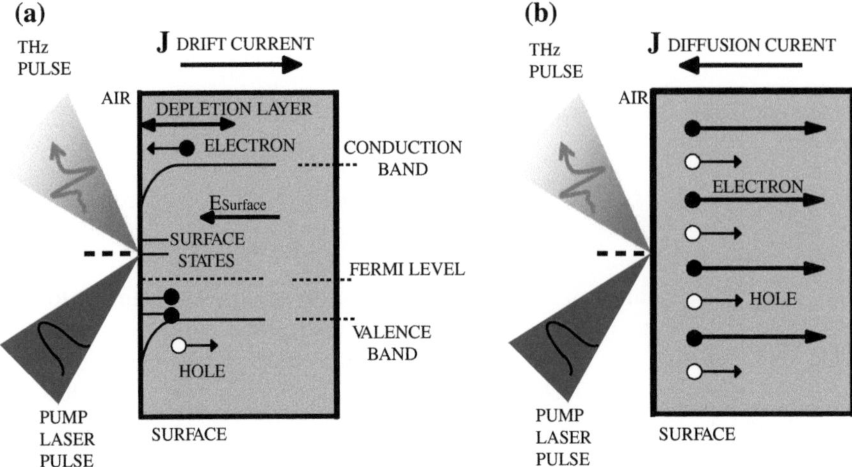

Fig. 2.4 Schematic diagrams of two types of current surge models of THz emission: **a** surface field effect, and **b** photo-Dember effect

In semiconductors with wide band gap such as GaAs ($E_g = 1.43$ eV), the surface states lie within the band gap of the semiconductor, and thus Fermi level pinning occurs, leading to band bending and the formation of a depletion region where the surface field exists. When an ultrafast laser pulse with photon energy greater than the band gap of the semiconductor material illuminates its surface, electron hole pairs are created. The electrons and holes are accelerated in the opposite directions under the influence of the surface depletion field, forming a surge current in the direction normal to the semiconductor surface. The far-field amplitude of the radiated THz electric field is proportional to the time derivative of the surge current. The direction and magnitude of the surface depletion field depends on the dopants or impurity species in the semiconductor, and the position of the surface states with respective to the bulk Fermi-level [3]. In general, the energy band is bent upward for n-type semiconductors and downward for p-type semiconductors. Therefore the built-in surface field in p-type semiconductors drives the photogenerated carriers (and thus the surge current) in the opposite direction to that in n-type semiconductors. This change in direction of the surge current results in polarity reversal of the emitted THz waveform between opposite semiconductor types. The amplitude of the radiated THz electric field in the far-field due to the surface field acceleration effect is given by [24]

$$E_{\text{far}} = \frac{Se^2 \mu_n G \left| N_D^+ - N_A^- \right|}{4\pi\epsilon_0^2 \epsilon_r c^3 \tau_r \alpha^2} \left[W\alpha - 1 + e^{-\alpha W} \right] \tag{2.54}$$

where, S is the laser focal spot size, τ is the laser pulse duration, r is the distance in the far field, G is the photocarrier generation rate, μ_n is the mobility of electrons, and ε_r is the permittivity of the semiconductor. Further, the magnitude of THz radiation due to surface field acceleration depends on the depletion width $W = ((2\varepsilon_r\varepsilon_0\Phi_S)/\rho)^{1/2}$, charge in the depletion region $\rho = q(p - n + N_D^+ - N_A^-) \approx q(N_D^+ - N_A^-)$, N_D^+ and N_A^- donor and acceptor concentrations respectively, and the absorption depth α.

The Photo-Dember field, on the other hand, is known to originate from the charge density gradient of photogenerated carriers near the surface. Since electrons have a higher mobility than holes, a space charge distribution is formed as electrons diffuse into the bulk material much faster than the holes. In the far-field region the amplitude of the radiated THz electric field is given by the time derivative of this diffusion current. The diffusion current due to the photo-Dember effect after photoexcitation of carriers near the semiconductor surface is shown in Fig. 2.4b. Since the mobility of electrons is always greater than that of the holes, the direction of diffusion current is the same irrespective of the semiconductor doping (n or p). Therefore, the polarity of the THz waveform due to the photo-Dember effect will be the same for both n and p-type semiconductors. The amplitude of the radiated THz electric field in the far-field due to the photo-Dember effect is given by [24]

$$E_{\text{far}} = \frac{S}{4\pi\varepsilon_0 c^2 \tau_r} \frac{k_B T_e \mu_n}{(1+b)} \left[\frac{b(n_b - p_b)}{1+b} \ln\left(1 + \frac{G(1+b)}{p_b + n_b}\right) - G \right] \tag{2.55}$$

where, n_b and p_b are electron and hole concentrations respectively, $b = \mu_n/\mu_h$ is the ratio of electron and hole mobilities, and T_e electron temperature. Indeed, the experimental evidence of lack of polarity reversal [5, 38] and a strong temperature dependence of the THz emission process [39] in narrow band gap semiconductors support photo-Dember effect as the dominant mechanism.

2.3.2 Optical Rectification

The THz optical rectification process can be considered as a nearly degenerate difference frequency mixing process between two different frequency components in a spectrally broadened laser pulse inside a nonlinear medium. When a pulsed light beam, containing a broad frequency spectrum which is determined by the shape and the duration of the optical pulse, is incident on a nonlinear crystal, the nonlinear interaction between two frequency components will induce a polarization and radiate electromagnetic waves at their beat frequency [2, 3]. Rectification can be a second order (difference frequency generation) or higher order nonlinear optical process depending on the optical power density [40]. The THz radiation field, generated by the optical rectification process is proportional to the second order nonlinear polarization (in the near field), and is described by the following equations [3]:

$$P(t) = \frac{1}{2\pi} \int\limits_{-\infty}^{\infty} P^{(2)}(\Omega)e^{-i\Omega t}d\Omega \tag{2.56}$$

$$P_i^{(2)}(\Omega) = \sum_{j,k} \varepsilon_0 \chi_{ijk}^{(2)}(\Omega = \omega_1 - \omega_2) \cdot \int\limits_{-\infty}^{\infty} E_j(\omega_1 = \Omega + \omega_2)E_k(\omega_2)d\omega_2 \tag{2.57}$$

where $\chi_{ijk}^{(2)}$ is the second-order non linear susceptibility tensor for a difference frequency, $\Omega = \omega_1 - \omega_2$, and $E_j(\omega_1)$ and $E_k(\omega_2)$ are the amplitude spectral components of the pump laser in the j and k directions respectively. Here, $i, j,$ and k are the dummy indices for the x, y, and z directions in the crystallographic axis system. In the far field, the observed THz field, $\mathbf{E}_{far}$, is proportional to the second time derivative of the nonlinear polarization $\mathbf{P}(t)$ and is given by [3]

$$\mathbf{E}_{far}(t) \propto \frac{\partial^2}{\partial t^2}\mathbf{P}(t). \tag{2.58}$$

The spectral amplitude $E_{far}(\Omega)$ at frequency Ω is given by

$$E_{far}(\Omega) \propto \Omega^2 \mathbf{n} \cdot \mathbf{P}(\Omega) \tag{2.59}$$

where $\mathbf{n}$ is the unit vector normal to the direction of observation. The experimental evidence of THz optical rectification is manifested in the strong dependence of emitted THz radiation intensity on crystal orientation relative to the pump laser polarization [25, 41–43]. By rotating the sample about its surface normal, the relative contribution from the azimuthal angle dependent difference frequency mixing component can be estimated.

References

1. Heyman, J., Coates, N., Reinhardt, A., Stresser, G.: Diffusion and drift in terahertz emission at GaAs surfaces. Appl. Phys. Lett. **83**, 5476 (2003)
2. Wilke, I., Sengupta, S.: Nonlinear optical techniques for terahertz pulse generation and detection–optical rectification and electrooptic sampling. In: Dexheimer, S.L. (ed.) Terahertz Spectroscopy: Principles and Applications. Optical Science and Engineering, vol. 131, 41. CRC Press, Boca Raton (2007)
3. Gu, P., Tani, M.: Terahertz radiation from semiconductor surfaces. In: Sakai, K. (ed.) Terahertz Optoelectronics. Topics in Applied Physics, vol. 97, 63. Springer, Berlin (2005)
4. Auston, D.H., Zhang, X.-C.: Optoelectronic measurement of semiconductor surfaces and interfaces with femtosecond optics. J. Appl. Phys. **71**, 326 (1992)
5. Gu, P., Tani, M., Kono, S., Sakai, K., Zhang, X.-C.: Study of terahertz radiation from InAs and InSb. J. Appl. Phys. **91**, 5533 (2002)
6. Darrow, J.T., Zhang, X.-C., Auston, D.H., Morse, J.D.: Saturation properties of large aperture photoconducting antennas. IEEE J. Quantum Electron. **28**, 1607 (1992)
7. Darrow, J.T., Auston, D.H., Morse, J.D.: Large-aperture photoconducting antennas excited by high optical fluences. In: Proceedings of SPIE, vol. 1861, 186 (1993)

8. Benicewicz, P.K., Roberts, J.P., Taylor, A.J.: Scaling of terahertz radiation from large-aperture biased photoconductors. J. Opt. Soc. Am. B **11**, 2533 (1994)

9. Stone, M.R., Naftaly, M., Miles, R.E., Fletcher, J.R., Steenson, D.P.: Electrical and radiation characteristics of semilarge photoconductive terahertz emitters. IEEE Trans. Microw. Theory Tech. **52**, 2420 (2004)

10. Goldberg, Y.A., Shmidt, N.M.: Gallium indium arsenide. In: Levinshtein, M., Rumayantsev, S., Shur, M.S. (eds.) Handbook Series on Semiconductor Parameters, vol. 2, 62. World Scientific, Singapore (1999)

11. Dutta, P.S.: III–V Ternary bulk substrate growth technology: a review. J. Crystal. Growth **275**, 106 (2005)

12. Künzel, H., Bottcher, J., Gibis, R., Urmann, G.: Material properties of $Ga_{0.47}In_{0.53}As$ grown on InP by low-temperature molecular beam epitaxy. Appl. Phys. Lett. **61**, 1347 (1992)

13. Sukhotin, M., Brown, E.R., Gossard, A.C., Driscoll, D., Hanson, M., Maker, P., Muller, R.: Photomixing and photoconductor measurements on ErAes/InGaAs at 1.55 μm. Appl. Phys. Lett. **82**, 3116 (2003)

14. Baker, C., Gregory, I.S., Tribe, W.R., Bradley, I.V., Evans, M.J., Withers, M., Taday, P.F., Wallace, V.P., Linfield, E.H., Davies, A.G., Missous, M.: THz-pulse imaging with 1.06 μm laser excitation. Appl. Phys. Lett. **83**, 4113 (2003)

15. Baker, C., Gregory, I.S., Tribe, W.R., Evans, M.J., Missous, M., Linfield, E.H.: Continuous-wave terahertz photomixing in low-temperature InGaAs. THz Technology, Ultrafast Systems and Measurements, IEEE, **367** (2004)

16. Chimot, N., Mangeney, J., Joulaud, L., Crozat, P., Bernas, H., Blary, K., Lampin, J.F.: Terahertz radiation from heavy-ion-irradiated $In_{0.53}Ga_{0.47}As$ photoconductive antenna excited at 1.55 μm. Appl. Phys. Lett. **87**, 193510 (2005)

17. Mangeney, J., Joulaud, L., Crozat, P., Lourtioz, J.-M., Decobert, J.: Ultrafast response (2.2 ps) of ion-irradiated InGaAs photoconductive switch at 1.55 μm. Appl. Phys. Lett. **83**, 5551 (2003)

18. Suzuki, M., Tonouchi, M.: Fe-implanted InGaAs emitters for 1.56 mm wavelength excitation. Appl. Phys. Lett. **86**, 051104-1 (2005)

19. Suzuki, M., Ohtake, H., Hirosumi, H., Tonouchi, M.: THz radiation from In[x]Ga[1-x]As (x = 1, 0.53, 0.60) excited by femtosecond lasers at wavelengths of 1560, 1050 and 780 nm. In: The Joint 30th International Conference on Infrared and Millimeter Waves, vol. 1, 328 (2005)

20. Takazato, A., Kamakura, M., Matsui, T., Kitagawa, J., Kadoya, Y.: Terahertz wave emission and detection using photoconductive antennas made on low-temperature-grown InGaAs with 1.56 μm pulse excitation. Appl. Phys. Lett. **91**, 011102-1 (2007)

21. Nemec, H., Pashkin, A., Kuzel, P., Khazan, M., Schnull, S., Wilke, I.: Carrier dynamics in low-temperature grown GaAs studied by THz emission spectroscopy. J. Appl. Phys. **90**, 1303 (2001)

22. Data obtained from vendor, MTI Corp. http://www.mtixtl.com/

23. Papastimatiou, M.J., Papaioannou, G. J.: Recombination mechanism and carrier lifetimes in semi-insulating GaAs:Cr. J. Appl. Phys. **68** (1990)

24. Ko, Y.: Electrical, optical, and THz emission studies of GaxIn1-xAs bulk crystals. Department of Electrical, Computer, and Systems Engineering, Doctoral Thesis, Rensselaer Polytechnic Institute, Troy, NY, 12180 (2007)

25. Ko, Y., Sengupta, S., Tomasulo, S., Dutta, P., Wilke, I.: Emission of terahertz-frequency electromagnetic radiation from bulk $Ga_xIn_{1-x}As$ crystals. Phys. Rev. B **78**, 035201 (2008)

26. Zhao, G., Schouten, R.N., van der Valk, N., Wenckebach, W.T., Planken, P.C.M.: Design and performance of a THz emission and detection setup based on semi-insulating GaAs emitter. Rev. Sci. Instrum. **73**, 1715 (2002)

27. Griffiths, D.J.: Introduction to Electrodynamics, 2nd edn, p. 284. Prentice-Hall of India Private Ltd., New Delhi (1994)

28. Griffiths, D.J.: Introduction to Electrodynamics, 2nd edn, p. 35. Prentice-Hall of India Private Ltd., New Delhi (1994)

29. Jackson, J.D.: Classical Electrodynamics, 2nd edn, p. 190. Wiley, New York (1975)

30. Kraus, J.D.: Antennas, 2nd edn, p. 207. McGraw-Hill, New York (1988)
31. Zhang, X.-C., Auston, D.H.: Optically induced THz electromagnetic radiation from planar photoconducting structures. J. Electromagn. Waves Appl. **6**, 85 (1992)
32. Darrow, J.T., Zhang, X.-C., Auston, D.H.: Power scaling of large aperture photoconducting antennas. Appl. Phys. Lett. **58**, 25 (1991)
33. Hu, B.B., Darrow, J.T., Zhang, X.-C., Auston, D.H., Smith, P.R.: Optically steerable photoconducting antennas. Appl. Phys. Lett. **56**, 886 (1990)
34. Darrow, J.T., Hu, B.B., Zhang, X.-C., Auston, D.H.: Subpicosecond electromagnetic pulses from large-aperture photoconducting antennas. Opt. Lett. **15**, 323 (1990)
35. NSM archive, Physical properties of Semiconductors. $Ga_xIn_{1-x}As$. [Online] [Cited: March 27, 2009.] http://www.ioffe.ru/SVA/NSM/Semicond/GaInAs/
36. Jackson, J.D.: Classical Electrodynamics, 2nd edn. Wiley, New York (1975). Chapter 6
37. Griffiths, D.J.: Introduction to electrodynamics, 2nd edn, p. 358. Prentice-Hall of India Private Ltd., New Delhi (1994)
38. Ascazubi, R., Schneider, C., Wilke, I., Pino, R., Dutta, P.: Enhanced THz emission from impurity compensated GaSb. Phys. Rev. B **72**, 045328 (2005)
39. Nakajima, M., Hangyo, M., Ohta, M., Miyazaki, H.: Polarity reversal of terahertz waves radiated from semi-insulating InP surfaces induced by temperature. Phys. Rev. B **67**, 195308 (2003)
40. Zhang, X.-C.: Terahertz wave imaging: horizons and hurdles. Phys. Med. Biol. **47**, 3667 (2002)
41. Xie, X., Xu, J., Zhang, X.-C.: Terahertz wave generation and detection from a CdTe crystal characterized by different excitation wavelengths. Opt. Lett. **31**, 978 (2006)
42. Rice, A., et al.: Terahertz Optical Rectification from <110> zinc-blende crystals. Appl. Phys. Lett. **64**, 1324 (1994)
43. Jin, Y.H., Zhang, X.-C.: Terahertz optical rectification. J. Nonlinear Opt. Phys. **4**, 459 (1995)

Chapter 3
Ultrafast Spectroscopy

In Chap. 2 we outlined the theory for generation of sub-picosecond THz pulses from semi-large aperture dc biased photoconducting (PC) antenna emitters fabricated on semiconductor substrates. Since the dynamic response of such devices relies on ultrafast transient currents, a PC substrate with short carrier lifetime and high mobility is desired. Short carrier lifetimes also prevent the buildup of persistent photoconductivity between consecutive laser pulses and reduce device heating due to thermal runaway.

A common experimental technique often employed for determining carrier lifetimes is the femtosecond optical pump-probe reflectivity measurement [1]. In its simplest form, the pump-probe reflectivity measurement involves the excitation of the semiconductor sample under investigation by one pulse train (pump). The change in refractive index induced by the generation of free carriers by the pump beam is probed by a second pulse train (probe) which is suitably delayed with respect to the pump. The change in reflectivity ΔR between the pump ON and OFF conditions is given by:

$$\Delta R = R(n', \hbar\omega) - R(n, \hbar\omega) \tag{3.1}$$

where n & n' are real parts of indices of refraction before and after the pump ON condition respectively and Δn is the change in the real part of the refractive index induced in the sample by the pump beam, and $\hbar\omega$ is the photon energy of the laser. The value of n' is given by:

$$n' = n + \Delta n \tag{3.2}$$

It can be shown that the reflectivity change ΔR depends linearly on the change in real index of refraction Δn as [2]:

$$\frac{\Delta R}{R} = \frac{4\Delta n}{(n^2 - 1)} \tag{3.3}$$

S. Sengupta, *Characterization of Terahertz Emission from High Resistivity Fe-doped Bulk Ga$_{0.69}$In$_{0.31}$As Based Photoconducting Antennas*, Springer Theses, DOI: 10.1007/978-1-4419-8198-1_3, © Springer Science+Business Media, LLC 2011

The change in real index of refraction (Δn) is related to the change in the imaginary index of refraction (Δk) by the Kramers–Krönig relation [3]

$$\Delta n = \frac{2\hbar}{\pi} P \int_{0}^{\infty} \frac{\omega' \Delta k(E)}{E^2 - (\hbar\omega)^2} \, \mathrm{d}E = \frac{\hbar c}{\pi} P \int_{0}^{\infty} \frac{\Delta\alpha(E)}{E^2 - (\hbar\omega)^2} \, \mathrm{d}E \qquad (3.4)$$

where P denotes the Cauchy principal value of the integral, and $\Delta\alpha$ is the absorption change which is related to the change in the complex index of refraction by the equation

$$\omega' \Delta k(E) = \frac{c\Delta\alpha(E)}{2} \qquad (3.5)$$

From Eqs. 3.3–3.5, we can see that the change in the pump-probe reflectivity signal of an optically excited semiconductor is explained by the variation in the real index of refraction through several effects like the absorption change due to band filling (BF), bandgap renormalization (BGR), free carrier absorption (FCA) and carrier recombination [4, 5]. In this work, the quantitative analysis is limited to the FCA effect. The change in the index of refraction and thus, the change in the transient reflectivity signal can be modeled through an increase in the plasma frequency given by [2, 6]

$$\frac{\Delta R}{R} = \frac{4n}{(n^2 - 1)} \left(\frac{2\pi e^2}{\varepsilon m^* \omega^2} \right) \Delta n_{\mathrm{pc}} \qquad (3.6)$$

where e is the electron charge, ε is the dielectric permittivity of the semiconductor, m^* is the conduction band effective mass, ω is the plasma frequency, and Δn_{pc} is the photocarrier density.

For our experimental purpose we assume that the carriers rapidly form a hot thermalized distribution following the excitation of the semiconductor sample by the pump pulse. This assumption is based on the fact that we conduct our experiments at high excitation densities that favor the quick thermalization of hot carriers via carrier-carrier scattering. The cooling of hot carriers occurs via various phonon scattering mechanism among which scattering with LO phonons is the most dominant. In addition to carrier energy relaxation or cooling, it is also required to consider effects of reduction in carrier density due to both spatial and temporal processes such as, diffusion, recombination and capture by traps or defects. The time evolution of the pump-probe reflectivity signal is a result of the time dependant carrier energy distribution and carrier density in the conduction band.

Several recombination processes such as non-radiative recombination, interband radiative recombination and Auger recombination may take place in a direct band gap semiconductor. Localized recombination centers such as impurities, physical defects and surface defects are responsible for non-radiative recombination [7]. The non-radiative recombination lifetime τ_{nr} can be expressed through Shockley–Read–Hall theory [6] as:

$$\tau_{nr} = \frac{1}{N_{def}\sigma_{def}\upsilon_{th}} \tag{3.7}$$

where N_{def} is the density of defects, σ_{def} is the defect cross section and υ_{th} is the thermal velocity of the carriers. From this equation it is evident that at a constant carrier temperature, the non-radiative recombination lifetime depends only on the material parameters (such as defect density and defect cross section) and is independent of the photocarrier density.

Interband radiative recombination is proportional to carrier concentration and in the case of doped semiconductors the recombination is limited by minority carrier concentration. The recombination lifetime [7] is defined as:

$$\tau_{rad} = \frac{B_{rad}}{n_c} \tag{3.8}$$

where n_c is the minority carrier concentration and B_{rad} is the proportionality constant.

The recombination lifetime for Auger recombination has a higher order dependence on carrier concentration since it involves the transfer of electron–hole pair energy to a third carrier [8]. In summary, the recombination rates are expressed as:

$$\frac{1}{\tau} = \frac{1}{\tau_{nr}} + \frac{n_b + \Delta n_{pc}}{B_{rad}} + \frac{\left(n_b + \Delta n_{pc}\right)^2}{C_{Auger}} \tag{3.9}$$

This equation is important because in pump-probe reflectivity experiments the photo injected carrier density (Δn_{pc}) is high compared to the native carrier density (n_b). By observing the lifetime dependence on the photoinjection density, it is possible to distinguish among these recombination mechanisms. In what follows in Sect. 5.1 in Chap. 5, we will incorporate all the aforementioned effects in interpreting the temporal evolution of the pump-probe reflectivity of $Ga_xIn_{1-x}As$.

References

1. Shah, J.: Ultrafast spectroscopy of semiconductors and semiconductor nanostructures. In: Springer Series in Solid State Sciences, vol. 115, Chap. 1. Springer, Heidelberg (1996)
2. Ascazubi, R.: THz Emission Spectroscopy of Narrow bandgap semiconductors. Department of Physics, Applied Physics, and Astronomy, Doctoral Thesis, Rensselaer Polytechnic Institute, Troy, NY (2005)
3. Boyd, R.: Nonlinear Optics, 2nd edn, Chap. 1. Academic Press, San-Diego (2003)
4. Prabhu, S.S., Vengurlekar, A.S.: Dynamics of the pump-probe reflectivity spectra in GaAs and GaN. J Appl. Phys. **95**, 7803 (2004)
5. Bennet, B.R., Soref, R.A., Del Alamo, J.A.: Carrier-induced change in refractive index of InP, GaAs, and InGaAsP. IEEE J. Quantum Electron. **26**, 113 (1990)

6. Korn, T., Franke-Wiekhorst, A., SchnÄull, S., Wilke, I.: Characterization of nanometer As-clusters in low-temperature grown GaAs by transient reflectivity measurements. J. Appl. Phys. **91**, 2333 (2002)
7. Sze, S.: Physics of Semiconductor Devices. Wiley, New York (1991)
8. Chen, F., Cartwright, A., Lu, H., Schaff, W.: Ultrafast carrier dynamics in InN epilayers. J. Cryst. Growth **269**, 10 (2004)

Chapter 4
Experimental Techniques

In this chapter we review in detail the experimental arrangements employed for the work described in this thesis. The remainder of this chapter is divided into four sections. Sects. 4.1 and 4.2 describe in details the THz time-domain spectroscopy and THz power measurement setups. In Sect. 4.3 we describe the ultrafast spectroscopy setup employed for investigating the carrier dynamics in bulk $Ga_xIn_{1-x}As$ crystals. Section 4.4 describes the details of bulk crystal growth of high-resistivity $Ga_{0.69}In_{0.31}As$ followed by a brief account on wafer processing and electrical characterization techniques.

4.1 Terahertz Time-Domain Spectroscopy Setup

The THz time-domain spectroscopy system used for the work described in this thesis is a typical pump-probe arrangement [1]. A schematic diagram of the setup has been depicted in Fig. 4.1. A commercial diode pumped Ti:S laser delivering pulses with 80–130 fs duration at a center wavelength of 800 nm is used to photo-excite the photoconductive (PC) antennas formed by attaching two planar electrodes on a $Ga_{0.69}In_{0.31}As$ substrate. The laser repetition rate is 82 MHz and the maximum output power is 650 mW which results in a pulse energy of ~ 8 nJ for the present configuration. An uncoated beam-splitter is used to split the laser beam into pump and probe beam with a typical power ratio of 95% on the pump beam and 5% on the probe beam. Emission of THz radiation is achieved by exposing the surface of the PC antenna to the pump beam. An external electric field up to 10 kV/cm is applied to enhance the THz radiation emission from the $Ga_{0.69}In_{0.31}As$ PC antenna. The angle between the pump laser beam and the emitter surface normal is set to Brewster's angle. In some cases the angle is set to $45°$ for the purpose of convenient alignment. A half wave ($\lambda/2$) plate (not shown in diagram) is used to rotate the polarization of the pump beam parallel to the plane of

S. Sengupta, *Characterization of Terahertz Emission from High Resistivity*
Fe-doped Bulk Ga$_{0.69}$In$_{0.31}$As Based Photoconducting Antennas, Springer Theses,
DOI: 10.1007/978-1-4419-8198-1_4, © Springer Science+Business Media, LLC 2011

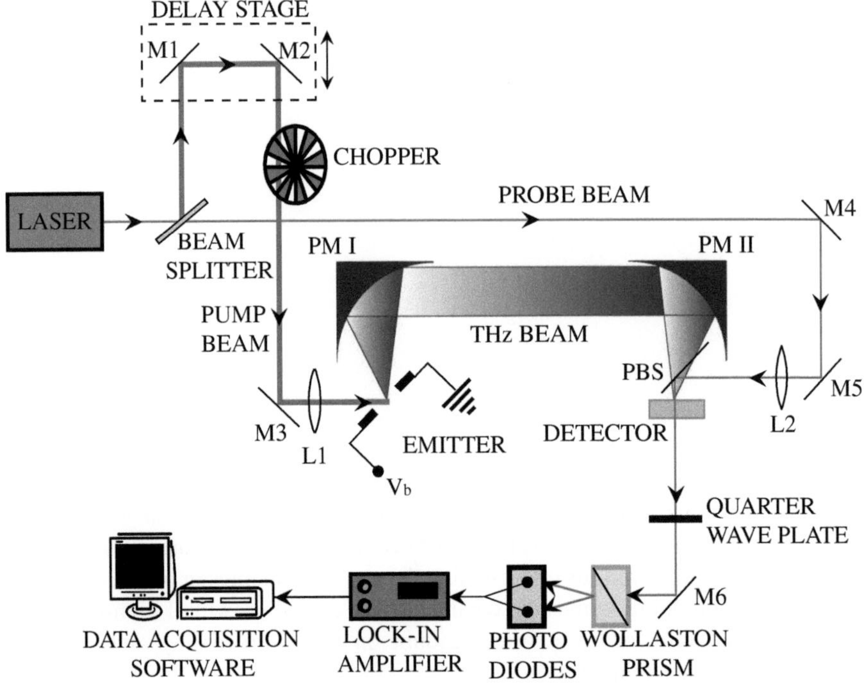

Fig. 4.1 Schematic diagram depicting the experimental arrangements of a typical THz time-domain spectroscopy system. (M1–M7 are optical beam steering mirrors, PMI and PMII are off-axis parabolic THz beam steering mirrors, L1 and L2 are lenses, PBS is pellicle beam splitter) [5]

incidence in order to minimize Fresnel losses at the sample surface and maximize the injected photocarrier density into the material. The pump beam is slightly focused to a spot size of 1 mm diameter on the surface of the PC antenna which releases a sub-picosecond pulse of THz radiation in response to the incident femtosecond near infrared laser pulse. The generated THz radiation is focused onto a detector using two off-axis gold coated parabolic mirrors. Pulsed THz radiation is detected through electro-optic sampling [2–4] using a 1 mm thick < 110 > ZnTe crystal. The probe beam gates the detector, whose response is proportional to the amplitude and sign of the electric field of the THz pulse. By changing the time delay between the pump and the probe beams by means of an optical delay stage, the entire time profile of the THz transient can be traced.

Electro-optic detection of THz transients is possible when the THz radiation pulse and the probe beams coincide in a co-propagating geometry inside the electro-optic crystal. A pellicle beam splitter is put in the THz beam line for this purpose. As the THz pulse and probe beam co-propagate through the electro-optic crystal, a phase modulation is induced on the probe beam which depends on the electric field of the THz radiation [5].

The phase modulation of the probe beam is analyzed by a quarter wave ($\lambda/4$) plate and the beam is then split into two beams of orthogonal polarizations by a Wollaston prism. At this point, the phase modulation of the probe beam is converted to an intensity modulation of the two orthogonal polarizations of the probe beam, which are then steered into a pair of balanced photodiodes. A lock-in amplifier subsequently detects the difference in probe laser light intensities measured by the photodiodes.

The optimum condition for electro-optic sampling is met when the optical and THz pulses travel with the same velocity inside the electro-optic crystal, that is, a perfect phase matching condition is met. For a medium with dispersion at optical frequencies, phase matching is achieved when the phase velocity of THz wave is equal to the velocity of optical pulse envelope (or the optical group velocity) [6]. However, the refractive indices for the near infrared laser frequency and THz frequency are generally not identical. Consequently, electromagnetic waves at THz and near infrared frequencies travel at slightly different speeds through the crystal. The efficiency of THz radiation detection using electro-optic sampling decreases if the mismatch between the velocities becomes too large. The distance over which the slight velocity mismatch can be tolerated is called the coherence length. As a result, efficient THz detection at a given frequency only occurs for crystals that are equal in thickness or thinner than the coherence length for this frequency.

If the phase matching problem is disregarded, the detected THz signal can be approximated as [7]:

$$\frac{\Delta I}{I} \alpha \, r \int\limits_{0}^{l} \int\limits_{-\infty}^{+\infty} T(\tau) E_{\text{THz}}(z, t - \tau) d\tau dz \qquad (4.1)$$

where I is the single diode current, ΔI is the difference in photocurrent in the diodes generated by the THz electric field, r is the electro-optic coefficient of the detector crystal, l is the crystal thickness, $T(\tau)$ is a function describing the time evolution of the laser pulse, and E_{THz} is the THz electric field. Equation 4.1 is a convolution where the result is the "average" of the electric field magnitude taken over the extent of the probe pulse. It is evident from this equation that the bandwidth of the system is limited by the laser pulse width. Equation 4.1 also implies that the detector signal is proportional to the THz electric field and the detector crystal thickness.

Thus, in electro-optic sampling the crystal thickness is a compromise between signal strength and bandwidth. The ZnTe crystal employed in this work has a thickness of 1 mm. This configuration has been found to yield a strong signal while it delivers a reasonable bandwidth. An optical chopper is used to modulate the pump beam, and the signal is detected with a lock-in amplifier.

It is noteworthy that both amplitude and phase information of the THz field is retained in electro-optic sampling since the sign of the electric field is also carried by the detected signal. The combination of optical pump-probe arrangement and the gated coherent detection process enables the system to achieve high signal–noise ratios of $\sim 10^3{:}1$.

4.2 Experimental Arrangement for Measurement of Average Terahertz Power

For the purpose of measuring the average power of the emitted THz radiation from $Ga_{0.69}In_{0.31}As$ PC antenna we employed the experimental arrangement depicted in Fig. 4.2.

The laser beam, focused to a spot size of 1 mm diameter, is used to photo-excite the $Ga_{0.69}In_{0.31}As$ PC antenna. The emitted THz radiation from the PC antenna is steered by two off-axis parabolic mirrors and focused onto a Golay cell detector. The window of the Golay cell is covered by a thin black poly-ethylene sheet in order to ensure that only the THz radiation passes through. The sensitivity of the Golay cell is 1.5×10^5 V/W and the noise equivalent power (NEP) is 10^{-10} W/Hz$^{1/2}$. The technical specifications for the Golay cell are provided in Appendix A. The laser beam was mechanically chopped at 11 Hz in order to accommodate the slow response time of the Golay cell and to operate within its optimum SNR regime. The excitation power from the Ti:S laser is modulated using a neutral density filter. All measurements are carried out in air and at room-temperature.

Figure 4.3a, b depict the semi-large aperture $Ga_{0.69}In_{0.31}As$ PC antennas employed in our experiments and the schematics of THz radiation emission from

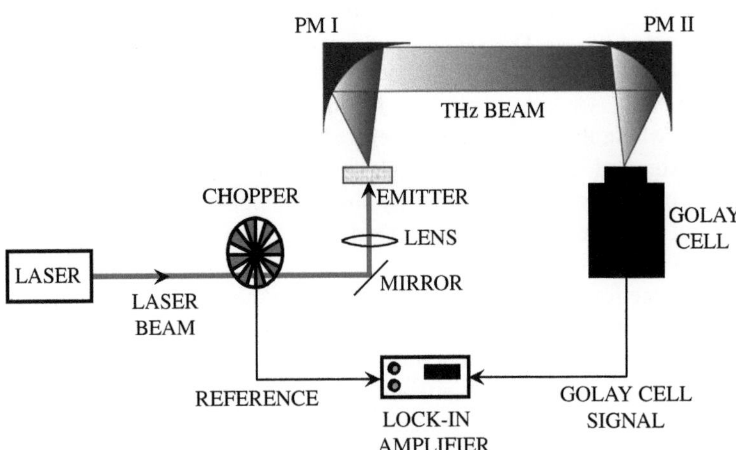

Fig. 4.2 Schematic diagram depicting the experimental arrangements for measuring THz power [8]

the same. The PC antennas are formed by painting two crescent shaped electrodes on the $Ga_{0.69}In_{0.31}As$ substrate with the smallest separation varying from 0.2 to 0.8 mm (Fig. 4.3a). Two large, polished, planar copper electrodes are attached to the silver paint electrodes using conductive silver epoxy at the separation of ~ 2 mm as shown in Fig. 4.3a, b. The silver paint and silver epoxy ensure good electrical and thermal contact between the electrodes and the semiconductor wafer's surface. This is important for the stable operation of the emitter since its overheating results in drop of its resistance and consequently makes higher voltage impossible to be applied. The silver paint electrodes are made about 1 mm broader than the contact area between the copper electrodes and the semiconductor substrate to ensure smooth voltage drop at the metal–semiconductor interface and prevent the erosion of the $Ga_{0.69}In_{0.31}As$ wafer due to the arc discharge through the air.

The high voltage bias applied to the copper electrodes of the PC antenna is provided by a Stanford Research Systems PS300 dc power supply. For the purpose of THz power measurement experiments the $Ga_{0.69}In_{0.31}As$ PC antenna is set at a normal geometry with respect to the incident laser beam as shown in Figs. 4.2 and 4.3b. The pump laser beam was focused between the electrodes to a slightly larger spot diameter (~ 1 mm) than the minimum electrode spacing to ensure uniform photoexcitation (Fig. 4.3a).

The emitted THz wave was coupled out of the opposite face of the $Ga_xIn_{1-x}As$ substrate using a 6 mm diameter hemispherical silicon lens as shown in Fig. 4.3b. For the purpose of the present experimental configurations passive cooling in air by heat dissipation from the large copper electrodes ensures thermally stable operation of the $Ga_xIn_{1-x}As$ PC antenna. However, for driving the PC antenna at higher excitation power and higher bias voltages it may be beneficial to incorporate active cooling mechanisms using a Peltier cooler to reduce surface heating effects. Cooling has been reported to produce an overall increase in THz power generated by this type of emitter [9].

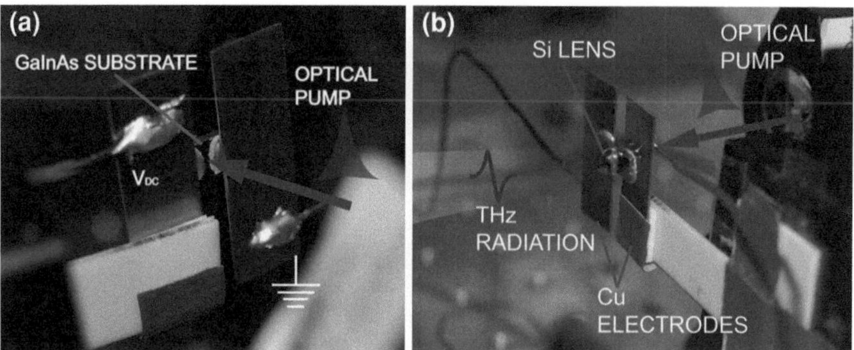

Fig. 4.3 a, b Semi-large PC antenna fabricated on $Ga_xIn_{1-x}As$ substrate. A hemispherical Si lens ensures the efficient coupling and collimation of the THz radiation onto free space

4.3 Ultrafast Spectroscopy Setup

Transient photo-reflection measurements have been investigated using a typical optical pump-probe arrangement [1]. A simplified schematic of the experimental setup has been presented in Fig. 4.4. The samples are excited using a Ti:Sapphire laser producing 130 fs pulses at a center wavelength of 800 nm and 82 MHz repetition rate. The maximum laser power is 650 mW. The pump beam injects into the semiconductor sample a high density of photocarriers which induce a change in the refractive index of the material. This change is probed by a second pulse train (probe beam) which is reflected off the semiconductor surface and suitably delayed with respect to the pump beam.

The pump and the probe beam must spatially and temporally overlap at the semiconductor surface. Further the spot size of the probe beam is smaller than the pump beam to ensure a condition of uniform photoexcitation. Both beams are focused on the sample using a lens with a focal length of 200 mm. A 200 μm pinhole was used to achieve the spatial overlap between the pump and the probe beam as well as to estimate the focal spot size on the sample. The estimated photoexcitation density is $4.0 \times 10^{18}/\text{cm}^3$. All measurements have been performed at room temperature. The probe beam polarization is made orthogonal to that of the pump beam to avoid any unwanted interference effects or coherent artifacts. A half wave ($\lambda/2$) plate and a Glan–Thompson polarizer are used to improve the signal quality. In order to study the dependence of the photo-reflection signal on the pump power a variable density neutral filter is used. An optical chopper modulator used in conjunction with lock-in detection technique enabled the detection of reflectivity variations ($\Delta R/R$) as low as $\sim 10^{-5}$.

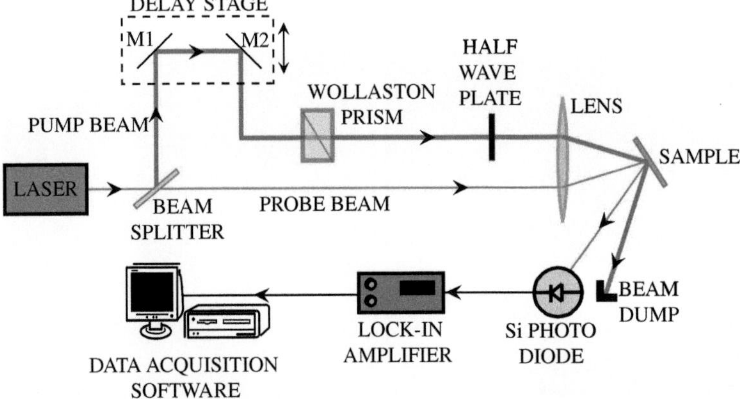

Fig. 4.4 Schematic diagram of ultrafast spectroscopy setup. An optical pump-probe arrangement is used to perform a time resolved measurement of the sample's reflectivity change

4.4 Bulk Crystal Growth of High Resistivity Ga$_x$In$_{1-x}$As: A Review [10]

In this section we briefly describe the experimental techniques employed for the growth of bulk Ga$_x$In$_{1-x}$As crystals. Figure 4.5 depicts the schematics of the crystal growth process. The growth of bulk ternary III–V semiconductors is extremely challenging since precise control of heat and mass transport in the high temperature melt during the crystal growth is necessary for obtaining high quality application worthy material [11]. For our experiments, a hybrid vertical Bridgman and gradient freezing directional solidification process was employed for the growth of high quality Ga$_x$In$_{1-x}$As bulk crystals with the uniform impurity doping concentration necessary for a high resistivity material.

The primary challenge faced during ternary crystal growth is the continuously changing composition along the growth direction due to the large separation between the solidus and liquidus curves in the phase diagram. Melt homogenization and solute feeding processes in conjunction with temperature gradient in the melt and solid is necessary for the successful growth of high quality compositionally homogeneous ternary crystals. Accelerated crucible rotation technique (ACRT) was used for melt homogenization during Ga$_x$In$_{1-x}$As crystal growth. In ACRT, the crucible is periodically accelerated and de-accelerated (around the growth axis) to promote efficient mixing of the melt. In addition to melt mixing, the shape of the melt-solid interface that decides the radial compositional profile in ternary crystals needs to be adjusted. A planar interface is an absolute necessity for

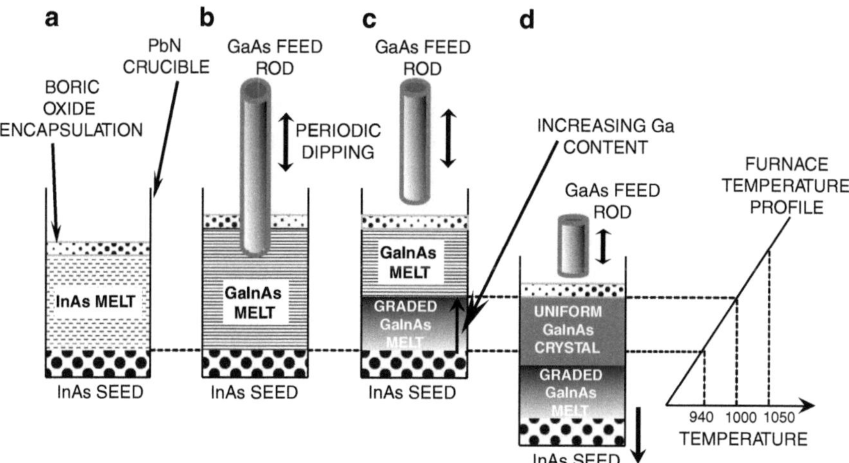

Fig. 4.5 Schematic diagram depicting the growth of a graded a homogeneous composition Ga$_x$In$_{1-x}$As crystal using vertical Bridgman and *vertical* and *horizontal* gradient freezing techniques. **a–d** The four stages of the crystal growth process are described in detail in the text. Accelerated crucible rotation is used during the entire experimental process

obtaining wafers with uniform alloy composition. An axial temperature gradient of 10–15°C/cm was used for achieving planar interface during growth. While optimized temperature gradient and ACRT provide compositional homogeneity in the radial direction (perpendicular to growth axis), melt replenishment is necessary for an axially homogeneous crystal. A periodic solute feeding process was developed for maintaining homogeneous melt compositions during the crystal growth.

The first step in ternary crystal growth is to generate a seed of a specific ternary composition starting from a binary seed. During this process, the melt composition needs to be continuously changed for grading the alloy composition of the crystal. The second step is to grow a homogeneous crystal of a specific composition. For the latter process, the melt needs to be maintained at a constant composition throughout the growth. In both steps precise control over the solute feeding process and melt composition is necessary. We now proceed to describe the growth of a homogeneous $Ga_xIn_{1-x}As$ crystal.

To start with, an undoped InAs seed crystal is placed at the bottom of the crucible along with undoped InAs polycrystalline charge (Fig. 4.5a). A GaAs polycrystalline feed is suspended from the top of the growth chamber. After heating and stabilizing the furnace to obtain a specific temperature gradient, the InAs charge melts along with a part of the InAs seed to obtain a melt-solid interface. At this point, the GaAs feed is lowered and allowed to touch the top of the InAs melt for a few seconds. The crucible is set to perform accelerated crucible rotation (ACRT) and a homogeneous GaInAs melt is prepared. The GaAs feed is lowered periodically to touch the GaInAs melt and the dissolved species is transported rapidly to the InAs seed interface as a result of ACRT melt mixing (Fig. 4.5b). Crystal growth is initiated in the crucible as a result of increasing level of the solute concentration at the solid–liquid interface with time. When the solute concentration in the melt near the seed interface reaches the liquidus composition, precipitation takes place and $Ga_xIn_{1-x}As$ starts growing on InAs.

As the GaAs feed is periodically lowered to touch the melt for few seconds, more solute is dissolved and transported to the growth interface and the crystal growth continues. The seed crystal grown by this method is compositionally graded along the growth axis with increasing gallium concentration in $Ga_{1-x}In_xAs$ and decreasing indium concentration as depicted in Fig. 4.5c. The rate of compositional grading is decided by the solidus temperature in the pseudo-binary phase diagram and the axial furnace temperature gradient. The axial composition in the seed crystal is graded till a desirable alloy composition is achieved and then a homogeneous composition crystal length is grown.

For the next step (axially uniform composition seed) of the crystal growth process, the crucible or the temperature gradient is translated along the length of the melt while the GaAs feed dissolution is continued by the periodic dipping (Fig. 4.5d). While the GaAs at the melt-solid interface is depleted by preferential incorporation in the crystal, it is replenished by the feed dissolution. Hence the melt-solid interface remains at the same position in the furnace till the entire melt solidified. To obtain high resistivity $Ga_{0.69}In_{0.31}As$, special efforts were made to grow GaAs polycrystalline feed uniformly doped with impurities such as Fe. These crystals were grown by multi-step directional solidification till the doping level in impurity content in the crystals reached in the range of 2–5 $\times$ 10^{16}/cm^3.

Figure 4.6a depicts a typical Ga$_x$In$_{1-x}$As ingot grown by the aforementioned technique. After crystal growth, the ingots were sliced into wafers of 1 mm thickness by a Low Speed Diamond Wheel Saw 660 from South Bay Technology. An Omnilap 2000 wafer polishing unit from South Bay Technology was used for lapping and polishing of the wafers. The lapped wafers were first polished with 1 μm alumina slurry on nylon pad followed by a second polishing with 0.01 μm alumina slurry on velvet pad to achieve mirror-like shining surfaces.

The band gap of the Ga$_x$In$_{1-x}$As crystals were obtained from FTIR measurements (Fig. 4.6b) while the resistivity, mobility and carrier concentrations were obtained from Hall Measurements performed using an HEM-2000 EGK Hall Measurement System from EGK Co.

A JEOL 733 electron probe microanalyzer (EPMA) was used to characterize the atomic compositions of the Ga$_x$In$_{1-x}$As crystals. For the as-grown Ga$_x$In$_{1-x}$As the band gap varies from 0.42 to 0.72 eV depending on the Ga composition. The mobility and carrier concentration vary from 16174 to 6,566 cm^2/V-s and 2.83 $\times$ 10^{16} to 5.30 $\times$ 10^{15}/cm^3 respectively. The as-grown Ga$_x$In$_{1-x}$As crystals have very low resistivity with the highest recorded value of $\sim$20 Ω-cm [12]. The band gap of high resistivity Fe-doped Ga$_{0.69}$In$_{0.31}$As is 1 eV. The resistivity, dc Hall mobility, and carrier concentration are 1.6 $\times$ 10^7 Ω cm, 2,395 cm^2/V-s, and 1.6 $\times$ 10^7/cm^3 respectively. The electrical and optical properties of Ga$_x$In$_{1-x}$As are summarized in Table 5.1 in Chap. 5. In the next section we will discuss in detail how these semiconductor physical parameters influence the THz radiation emission from Ga$_x$In$_{1-x}$As.

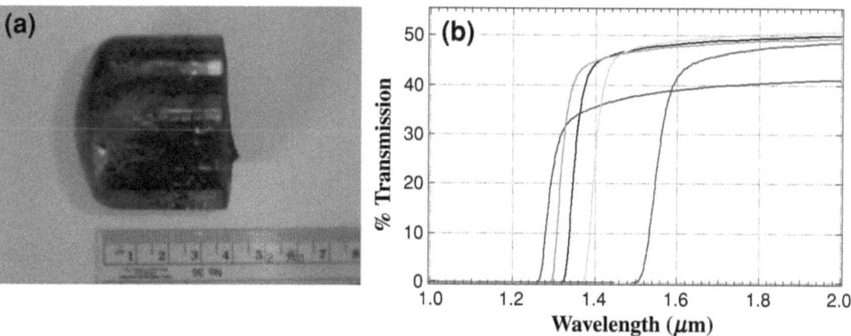

Fig. 4.6   a A bulk Ga$_x$In$_{1-x}$As crystal grown using vertical Bridgman and *vertical* and *horizontal* gradient freezing techniques with accelerated crucible rotation. b Transmission spectra of bulk Ga$_x$In$_{1-x}$As wafers with different band gaps (cut-off wavelengths corresponding to 1.3–3.1 μm are shown here)

References

1. Shah, J.: Ultrafast spectroscopy of semiconductors and semiconductor nanostructures. In: Springer Series in Solid State Sciences, vol. 115, Chap. 1. Springer, Heidelberg (1996)
2. Wilke, I., Sengupta, S.: Nonlinear optical techniques for terahertz pulse generation and detection—optical rectification and electrooptic sampling. In: Dexheimer, S.L. (ed.) Terahertz Spectroscopy: Principles and Applications, Optical Science and Engineering, vol. 131, p. 41. CRC Press, Boca Raton (2007)
3. Gallot, G., Grischkowsky, D.: Electrooptic detection of terahertz radiation. J. Opt. Soc. Am. B **16**, 1204 (1999)
4. Wu, Q., Litz, M., Zhang, X.-C.: Broadband detection capability of ZnTe electro-optic field sensors. Appl. Phys. Lett. **68**, 2924 (1996)
5. Wilke, I., Sengupta, S.: Nonlinear optical techniques for terahertz pulse generation and detection—optical rectification and electrooptic sampling. In: Dexheimer, S.L. (ed.) Terahertz Spectroscopy: Principles and Applications, Optical Science and Engineering, vol. 131. Taylor & Francis Group LLC—Books, London. Reproduced with permission of Taylor & Francis Group LLC—BOOKS in the format Other book via Copyright Clearance Center (2007)
6. Nahata, A., Weling, A.S., Heinz, T.F.: A wideband coherent terahertz spectroscopy system using optical rectification and electrooptic sampling. Appl. Phys. Lett. **69**, 2321 (1996)
7. Bakker, H., Cho, G., Kurz, H., Wu, Q., Zhang, X.: Distortion of THz pulses in electro-optic sampling. J. Opt. Soc. Am. B **15**, 1795 (1998)
8. Reprinted with permission from Sengupta, S., Wilke, I.S., Dutta, P.: Ultrafast carrier mobilities in high-resistivity iron-doped $Ga_{0.69}In_{0.31}As$ photoconducting antennas. Appl. Phys. Lett. **95**, 211102 (2009). Copyright 2009, American Institute of Physics
9. Zhao, G., Schouten, R.N., van der Valk, N., Wenckebach, W.T., Planken, P.C.M.: Design and performance of a THz emission and detection setup based on semi-insulating GaAs emitter. Rev. Sci. Instrum. **73**, 1715 (2002)
10. Communication with Professor Partha Dutta, Department of Electrical, Computer and Systems Engineering, Rensselaer Polytechnic Institute, 110 8th street, Troy, NY 12180
11. Dutta, P.S.: Crystal Growth Technology. In: Scheel, H.J., Capper, P. (eds.) Chap. 12. Wiley, Weinheim (2008)
12. Ko, Y.: Electrical, Optical, and THz emission studies of GaxIn1−xAs bulk crystals. Department of Electrical, Computer, and Systems Engineering, Doctoral Thesis, Rensselaer Polytechnic Institute, Troy (2007)

Chapter 5
Experimental Results

In Chap. 2 we reviewed the theoretical model of THz emission from semi-large aperture photoconducting (PC) antennas. Typically such antennas are fabricated on semi-insulating (SI) GaAs or low-temperature (LT) GaAs semiconductor substrates that exhibit high resistivity and high photocarrier mobility along with sub-picosecond carrier lifetimes. In the context of THz photoconducting antenna emitters, subpicosecond carrier lifetimes are important to prevent the buildup of persistent photoconductivity between consecutive laser pulses and reduce device heating. In Sect. 5.1 of this chapter, we investigate the carrier dynamics and estimate the carrier lifetime in as-grown $Ga_xIn_{1-x}As$ and high resistivity, Fe-doped $Ga_{0.69}In_{0.31}As$ using the transient photo-reflectivity measurement technique outlined in Sect. 4.3. The sub-picosecond carrier lifetimes observed in all of the $Ga_xIn_{1-x}As$ samples make them excellent candidates for THz PC antenna applications as we will see in the later sections of this chapter.

A typical THz time-domain spectroscopy system employs a PC antenna emitter photoexcited at 800 nm wavelength obtained from femtosecond Ti:S laser sources (refer to Fig. 1.1a in Chap. 1) for emission of THz frequency waves. However, since Ti:S lasers are large and complex and are limited in their output power, it is desirable to replace these lasers with compact, high-power, diode-pumped solid state lasers which operate at wavelengths between 1.1 and 1.5 μm. At the same time it is necessary to engineer novel narrow band gap materials which can be directly photoexcited at these wavelengths. $Ga_xIn_{1-x}As$ is an important material in this regards since its band gap can be varied from 0.36 to 1.42 eV by variation of Ga mole fraction x. Prior research [1, 2] on bare $Ga_xIn_{1-x}As$ crystals photoexcited at 800 nm has shown that the emitted THz radiation varies with Ga mole fraction x. In Sect. 5.2 we extend this work to investigate the THz emission from bare $Ga_xIn_{1-x}As$ crystals excited at 1.1 μm using a Yb:KYW laser (100–150 fs pulse duration, 85 MHz repetition rate) developed by Q-Peak Inc. Our experimental results indicate that it is possible to extract THz power of 867 nW from $Ga_{0.44}In_{0.56}As$ crystal when it is photoexcited by 800 mW of average laser power.

S. Sengupta, *Characterization of Terahertz Emission from High Resistivity Fe-doped Bulk Ga_0.69In_0.31As Based Photoconducting Antennas,* Springer Theses, DOI: 10.1007/978-1-4419-8198-1_5, © Springer Science+Business Media, LLC 2011

As a next step we proceed to fabricate $Ga_xIn_{1-x}As$ based semi-large PC antennas since it has been shown in literature [3] that these types of THz emitters have better efficiency for converting the incident laser power to THz power, compared to surface emitters. However, as-grown $Ga_xIn_{1-x}As$ has very low dark resistivity and is not capable of holding off the large bias fields required for the operation of semi-large aperture antennas. Using the process outlined in Sect. 4.4 in Chap. 4, we developed in collaboration with Professor Partha Dutta, $Ga_{0.69}In_{0.31}As$ crystals with high resistivity ($\sim 10^7\ \Omega$), high DC mobility ($2395\ cm^2/Vs$), and sub-picosecond carrier lifetimes. In Sect. 5.3 of this chapter, we study the THz emission from $Ga_{0.69}In_{0.31}As$ PC antennas photoexcited at 800 nm, as a function of applied bias voltage, incident laser power, and electrode gap spacing of the antenna. Our experimental results are in good agreement with the theoretical model of THz emission outlined in Chap. 2. From these results we are also able to estimate the ultrafast carrier mobility in $Ga_{0.69}In_{0.31}As$. This is an important parameter, since high carrier mobility in a material favors stronger THz emission. From the experimental data obtained on the THz emission from a $Ga_{0.69}In_{0.31}As$ PC antenna photoexcited at 800 nm, we are also able to simulate its performance when excited by Yb:doped laser at 1.1 μm. Our calculations predict that a $Ga_{0.69}In_{0.31}As$ PC antenna with 0.8 mm electrode spacing is capable of producing 51 μW of output THz power when photoexcited at 1.1 μm and at an average laser power of 1 W.

5.1 Ultrafast Recombination in $Ga_xIn_{1-x}As$

In this section we discuss, the ultrafast reflectivity measurements of as-grown and high resistivity Fe-doped bulk $Ga_{0.69}In_{0.31}As$. Numerous experiments of this type have been performed to date [4–8] and theoretical analysis attempted [9–12] in order to understand the fundamental carrier scattering processes in GaAs. However, experimental work on $Ga_xIn_{1-x}As$ has almost exclusively focused on few compositions of $Ga_xIn_{1-x}As$ quantum wells [13, 14] and thin films [15–19]. Since the physics of quantum wells is different from bulk materials it is necessary to conduct a detailed study of carrier dynamics in bulk $Ga_xIn_{1-x}As$. Further, an understanding of the fundamental carrier scattering processes in bulk $Ga_xIn_{1-x}As$ is also necessary from the standpoint of achieving ultrashort carrier lifetime in this material for possible photoconductive switching applications described in the previous section.

As mentioned in Chap. 3, the transient reflectivity of an optically excited semiconductor is explained by the variation of complex refractive index through several effects, such as the absorption change due to band filling (BF), band gap renormalization (BGR), free carrier absorption (FCA) and carrier recombination. The present interpretation of the temporal evolution of the transient reflectivity spectra of bulk $Ga_xIn_{1-x}As$ follows the calculations of BGR and BF effects outlined in [9]. It is assumed that the high excitation densities in our experiments favor the quick thermalization of hot carriers via carrier-carrier scattering and subsequent carrier cooling via phonon emission. This assumption is further supported by the experimental observation of rather long LO phonon scattering times (~ 150 fs) in Ga$_x$In$_{1-x}$As [20].

Figure 5.1a, b depict photo-reflection transients from as-grown and high resistivity Fe-doped bulk Ga$_{0.69}$In$_{0.31}$As respectively. The data was processed to remove any offsets and normalized with respect to the transient photo-reflectivity signal from high resistivity Fe doped Ga$_{0.69}$In$_{0.31}$As in Fig. 5.1b for the sake of comparison. It can be seen that the peak signal amplitude increases with increase in Ga composition. An increase in Ga composition results in an increase in band gap and this implies a shift towards near band gap excitations (given the constant value of excitation energy of 1.55 eV at a wavelength of 800 nm, obtained from a Ti:Sapphire laser). Since the carrier induced refractive index change increases rapidly near the band edge, and in fact, a resonance occurs, larger signals are expected for measurements closer to the band edge. This accounts for the increase in the peak reflectivity signal as the band gap gets closer to the excitation energy.

The temporal evolution of the reflectivity transients in both as-grown and high resistivity Fe-doped Ga$_{0.69}$In$_{0.31}$As samples are characterized by an initial rise in the signal followed by decay and a subsequent slow recovery resulting in a dip occurring ~ 1 ps after the initial rise. The decay of the signal does not occur at the decay rate of the carrier decay, but due to the combined effects of BGR and BF. If the BGR effect is made small, the decay of the pump-probe reflectivity signal is mainly governed by the carrier decay rate. The dip in the transient reflectivity spectra is a possible contribution from the BGR effect as outlined in [9]. The BF contribution to the pump-probe reflectivity signal is positive and the BGR contribution is negative [9]. Since the overall signature of the reflectivity transients in

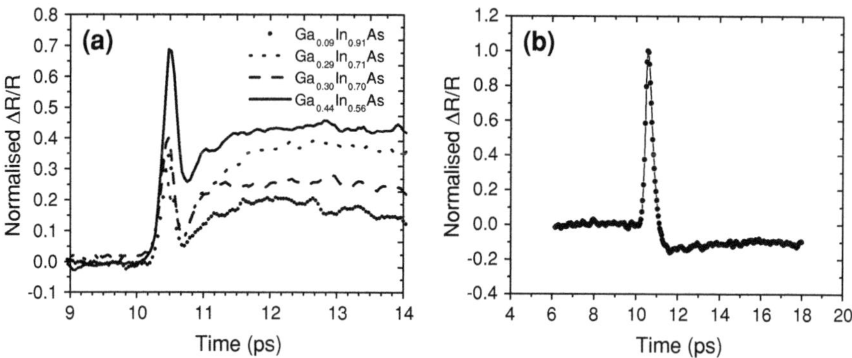

Fig. 5.1 **a** Recorded reflectivity transient for as-grown Ga$_x$In$_{1-x}$As samples W$_5$, W$_6$, W$_7$ and W$_8$ (for $x = 0.09, 0.29, 0.33, 0.44$ respectively). **b** Reflectivity transient for Fe doped high resistivity Ga$_{0.69}$In$_{0.31}$As. The reflectivity decay is fitted by a single exponential (exp $(-t/\tau)$) with $\tau \sim 0.3$ ps. The data have been normalized with respect to Ga$_{0.69}$In$_{0.31}$As [21]

Fig. 5.1a is also positive, it may be inferred that the BF effect offsets the BGR effect in as-grown narrow band gap $Ga_xIn_{1-x}As$.

It should also be noted that the photoexcitation of carriers for transient reflectivity experiments performed on as-grown narrow band gap $Ga_xIn_{1-x}As$ occurs far above the materials' band edge. Therefore intervalley scattering can play a significant role in determining the temporal evolution of the reflectivity change, especially on a sub-picosecond time scale. As carriers relax from L valley to the band edge in Γ valley, the change in absorption due to BF increases and offsets the decrease due to BGR and carrier decay. An intervalley transition time of ~ 2 ps from L to Γ valley has been previously reported for $Ga_{0.47}In_{0.53}As$ from pump-probe spectroscopy experiments [17]. Intervalley scattering events are estimated to be low in the case of high resistivity Fe doped $Ga_xIn_{1-x}As$ since in this case the photoexcitation occurs close to the band edge.

The carrier decay time due to carrier trapping and recombination in as-grown and Fe-doped high resistivity $Ga_{0.69}In_{0.31}As$ was obtained by fitting a single exponential to the reflectivity data presented in Fig. 5.1a, b. It must be kept in mind that the carrier decay time obtained thus does not reflect true carrier recombination lifetime and incorporates the complex combined effects of BF, BGR, and intervalley scattering events.

The carrier lifetime thus obtained ranges from ~ 62 fs for as-grown $Ga_{0.09}In_{0.91}$ As to ~ 306 fs for Fe-doped high resistivity $Ga_{0.69}In_{0.31}As$. The carrier lifetime for LT-$Ga_{0.47}In_{0.53}As$ estimated by Gupta et al. from pump-probe reflectivity spectra previously, was reported to be about ~ 2.5 ps [19, 22]. On the other hand, photoconductive switch response experiments performed on high resistivity lattice mismatched LT-$Ga_{0.75}In_{0.25}As$ on GaAs substrates yielded carrier lifetimes of 7.3 ps [22]. In contrast, the as-grown $Ga_xIn_{1-x}As$ samples used in this experimental work exhibit carrier lifetimes <1 ps. Since our as-grown $Ga_xIn_{1-x}As$ samples are polycrystalline in nature [1] it is possible that the large density of defects in the form of grain boundaries act as effective trapping and recombination centers thus reducing the carrier lifetimes. The sub-picosecond carrier lifetime observed in high resistivity Fe-doped $Ga_{0.69}In_{0.31}As$ is explained by carrier trapping by the deep level defects formed by the Fe atoms. Table 5.1 summarizes the semiconductor physical properties and carrier lifetimes for the $Ga_xIn_{1-x}As$ samples employed in this research.

Table 5.1 Carrier lifetimes and electrical properties for $Ga_xIn_{1-x}As$ samples at 300 K [21]

Sample	Ga (x)	In (1−x)	Band gap	Carrier concentration (cm^{-3})	Resistivity (Ω-cm)	Hall mobility (cm^2/Vs)	Carrier lifetime (fs)
W5	0.09	0.91	0.42	2.83×10^{16}		16174	61.8 ± 11
W6	0.29	0.71	0.59	9.49×10^{15}		11601	157.4 ± 59.8
W7	0.30	0.70	0.59	9.08×10^{15}	<20	10100	202.7 ± 162.8
W8	0.44	0.56	0.72	5.30×10^{15}		6566	131.1 ± 43.8
Fe-doped $Ga_{0.69}In_{0.31}As$	0.69	0.31	1.00	1.6×10^7	1.6×10^7	2395	305.9 ± 9.4

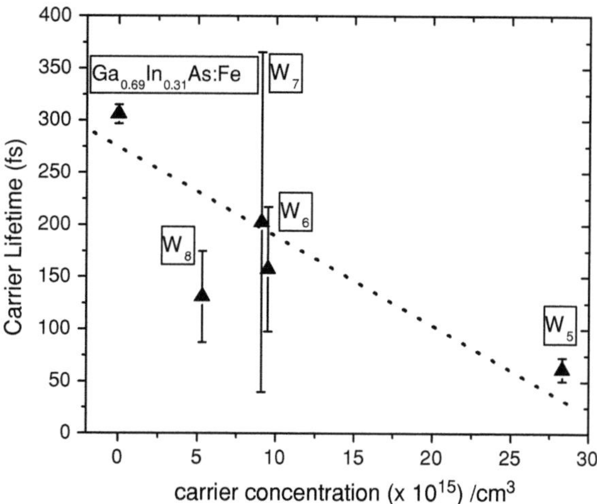

Fig. 5.2 Recombination lifetimes versus carrier concentrations. The carrier concentrations were measured using dc Hall measurements. For high resistivity Fe-doped Ga$_{0.69}$In$_{0.31}$As the carrier concentration is 1.6×10^7/cm^3. The dotted line is a guide for the eye

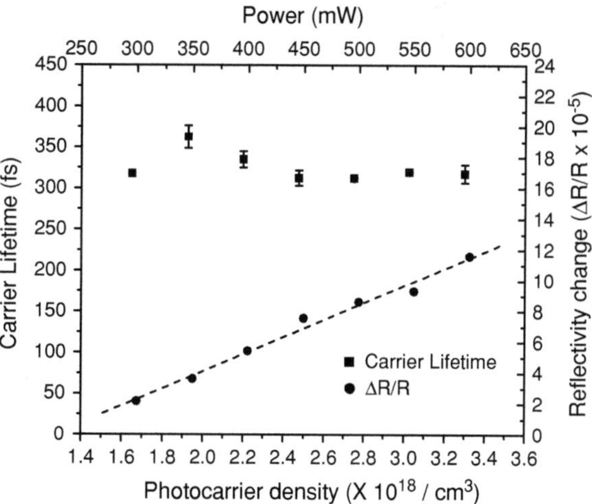

Fig. 5.3 The reflectivity change depends linearly on pump power, as expected from free carrier absorption. The carrier lifetime does not depend on the pump power, as expected for non-radiative recombination. The Ga$_{0.69}$In$_{0.31}$As sample has a carrier concentration of 1.6×10^7/cm^3 [21]

The carrier lifetimes versus carrier concentrations are illustrated in Fig. 5.2. From this figure we see a general decreasing trend in the carrier lifetime with increase in carrier concentration. The observation of an inverse relationship between carrier lifetime and carrier concentration has also been previously reported in narrow band gap semiconductors such as InN by Ascazubi et al. [23] and Chen et al. [24] and has been attributed to carrier recombination and trapping by the non-radiative recombination centers formed by donor-like defects or impurities [24].

In order to measure the nature of carrier recombination process in high resistivity Fe-doped Ga$_{0.69}$In$_{0.31}$As, reflectivity transients have been measured for various excitation powers. The observed peak signal amplitude versus pump power and carrier lifetime versus pump power has been depicted in Fig. 5.3.

The reflectivity change depicts a linear dependence on pump power. This is in agreement with free-carrier absorption. For this process the change in reflectivity is linearly proportional to the photocarrier density as shown in Eq. 3.6.

The carrier lifetime is independent of pump power within the error of our measurements. From Eq. 3.9 we can infer that non-radiative recombination is the dominant recombination mechanism in high resistivity Fe-doped $Ga_{0.69}In_{0.31}As$ at room temperature and for carrier densities up to $\sim 4 \times 10^{18}/cm^3$. Further, the contribution from Augur recombination process or radiative recombination processes can be assumed to be negligible since these are expected to have quadratic or higher order dependence on photo-injected carrier density.

In conclusion it may be said that sub-picosecond carrier lifetimes were observed in as-grown and high resistivity Fe-doped $Ga_{0.69}In_{0.31}As$. Non-radiative recombination was found to be the dominant carrier recombination mechanism in high resistivity Fe-doped $Ga_{0.69}In_{0.31}As$. Since the photoexcitation of carriers for all of the above experiments occur far above the $Ga_xIn_{1-x}As$ band edge it is likely that intervalley scattering events significantly affect the overall shape of the temporal evolution of the reflectivity data. Therefore to estimate the actual carrier lifetime in $Ga_xIn_{1-x}As$ on the sub-picosecond time scale it is preferable to photo-inject carriers at energy close to the band gap so that scattering to satellite valleys is completely eliminated. Nonetheless, our experiments provide valuable information regarding the fundamental carrier scattering processes in bulk $Ga_xIn_{1-x}As$ measured for several different Ga compositions. The observation of sub-picosecond carrier lifetime in high resistivity Fe-doped $Ga_{0.69}In_{0.31}As$ explains its superior performance as a semi-large aperture PC terahertz emitter as described in Sect. 5.3.

5.2 Terahertz Emission from Unbiased $Ga_xIn_{1-x}As$ Crystals Excited at 1.1 µm

In this section we discuss the results of the THz emission measurements performed on bare $Ga_xIn_{1-x}As$ crystals photoexcited at 1.1 µm. The THz emission experiments from unbiased $Ga_xIn_{1-x}As$ crystals were performed in collaboration with Q-Peak Inc. and Physical Science Corp. (PSI) using a setup similar to that shown in Fig. 4.2 in Chap. 4. However, in this case the laser beam was incident on the samples at an angle of 30° with respect to the surface normal (similar to that shown in Fig. 4.1). The detection of the average emitted THz power was performed by a deuterated triglycine sulfate (DTGS) pyroelectric detector which had a 3 mm diameter sensor area and about 3 V/W responsivity at ~ 70 µm wavelength for a ~ 100 Hz chopping frequency. A telescope lens assembly consisting of a positive and negative lens was used to vary the spot size of the pump beam and therefore the peak intensity of the laser pulse for the investigation of saturation effects. The laser spot has an elliptical shape at the sample surface.

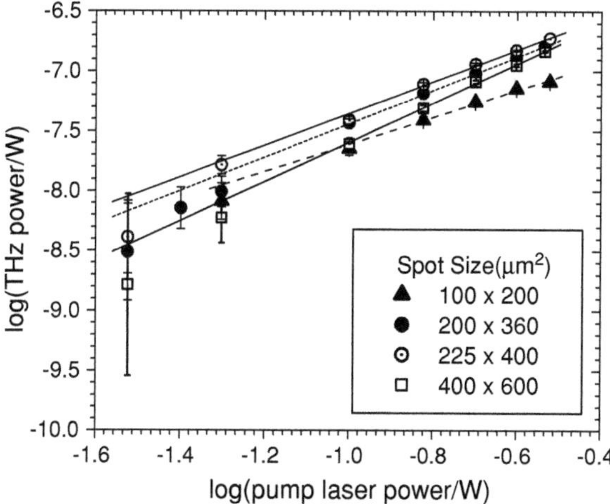

Fig. 5.4 Dependence of THz emission on pump beam spot size and power for Ga$_{0.44}$In$_{0.36}$As. The symbols are experimentally obtained data points and the solid, dashed, and dotted lines are linear fits of the data points. The samples were photoexcited by a Yb:KYW laser operating at a center wavelength of 1.1 µm and a maximum average power of 300 mW. The laser beam was incident on the sample at an angle of 30° with its surface-normal. The laser spot had an elliptical shape

Table 5.2 Slope of the linear fit of data from: (a) Fig. 5.1 and (b) Fig. 5.2

x (Ga)	Spot Size (µm²)	Slope
(a)		
0.44	100 × 200	1.13 ± 0.07
	200 × 360	1.40 ± 0.03
	225 × 400	1.33 ± 0.05
	400 × 600	1.64 ± 0.04

Spot Size (µm²)	x (Ga)	Slope
(b)		
225 × 400	0.44	1.33 ± 0.05
	0.30	1.85 ± 0.13
	0.29	1.06 ± 0.05
	0.09	1.47 ± 0.07

Figure 5.4 shows the dependence of emitted THz power on pump laser power for a range of laser beam spot sizes. The maximum average laser power in this case was 300 mW (150 fs pulse duration). From the slopes of the linear fits of these data points listed in Table 5.2a, we observe that the emitted THz power approaches a near-quadratic dependence on the incident pump laser power as we increase the spot size of the laser beam at the sample surface. For tighter focused beams and smaller spot sizes a sub-quadratic dependence of the emitted THz power on incident laser power is observed. Quadratic dependence of the emitted THz power

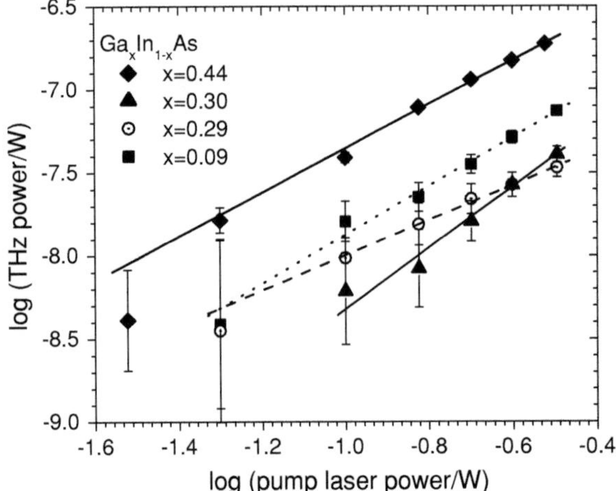

Fig. 5.5 Dependence of THz emission on pump power for varying Ga mole fraction. The symbols are experimentally obtained data points and the solid, dashed, and dotted lines are linear fits of the data points. The samples were photoexcited by a Yb:KYW laser operating at a center wavelength of 1.1 μm and a maximum average power of 300 mW. The laser beam was incident on the sample at an angle of 30° with its surface-normal. The laser spot had an elliptical shape

Table 5.3 Comparison of material properties and maximum emitted THz power for Ga$_x$In$_{1-x}$As emitters pumped with 300 mW

x (Ga)	Band gap (eV) [25]	DC Hall mobility (cm^2/Vs) [1, 2]	Emitted THz power (nW)
0.09	0.42	16174	74
0.29	0.59	11601	34
0.30	0.59	10100	41
0.44	0.72	6566	189

on laser power can be generally expected since the THz amplitude is proportional to the generated photocurrent, which is in turn proportional to the incident pump power. Such a behavior is also characteristic of optical rectification effect observed in Ga$_x$In$_{1-x}$As as outlined in Sect. 2.3.2 of Chap. 2. However, at higher laser powers, the screening of the incident radiation by photocarriers and thermally generated carriers may lead to sub-quadratic dependence of the emitted THz power on incident laser power.

Figure 5.5 shows the dependence of emitted THz power on incident laser power for varying Ga mole fractions in Ga$_x$In$_{1-x}$As emitters. These measurements were also carried out at a maximum average laser power of 300 mW and a laser beam spot size of 225 μm × 400 μm. Table 5.2b lists the slopes of the linear fit of the data in Table 5.3. As evident from the plots in Fig. 5.5, the emitted THz power generated from the other Ga compositions is much lesser when compared to the Ga$_{0.44}$In$_{0.66}$As sample and uncertainties in the fitted power dependence do not allow firm conclusions to be drawn regarding the dependence of emitted THz power on varying Ga compositions. The strongest emitted THz power (189 nW) was obtained for Ga mole fraction x = 0.44. This result is consistent with our prior calculations [1, 2] which predict that the THz emission from unbiased $Ga_xIn_{1-x}As$ due to surface field acceleration is maximized for Ga mole fraction x ∼ 0.45. A comparison of $Ga_xIn_{1-x}As$ material properties for different Ga mole fractions and the emitted THz power has been provided in Table 5.3.

Since $Ga_{0.44}In_{0.56}As$ shows the strongest THz emission compared to other Ga compositions, we chose this sample to perform a subsequent set of THz emission measurements at an increased average laser power of 800 mW and 180 fs pulse duration. Figure 5.6 depicts the results of this measurement, where the emitted THz power from $Ga_{0.44}In_{0.56}As$ is plotted as a function of the incident laser power for two different laser beam spot sizes.

Increasing the pump laser power resulted in an increase in the emitted THz power with a maximum emitted THz power of 867 nW when the source was pumped with 800 mW of laser power. From Fig. 5.6 it is evident that the THz power generated for the larger laser beam spot size (930 µm × 570 µm) has a nonlinear dependence on the pump laser power in comparison with that for the tighter focus (315 µm × 615 µm) which is almost linearly dependent on the pump power. The nonlinear dependence of THz power on the pump laser power may be attributed to the optical rectification effect in $Ga_xIn_{1-x}As$ as observed in an earlier

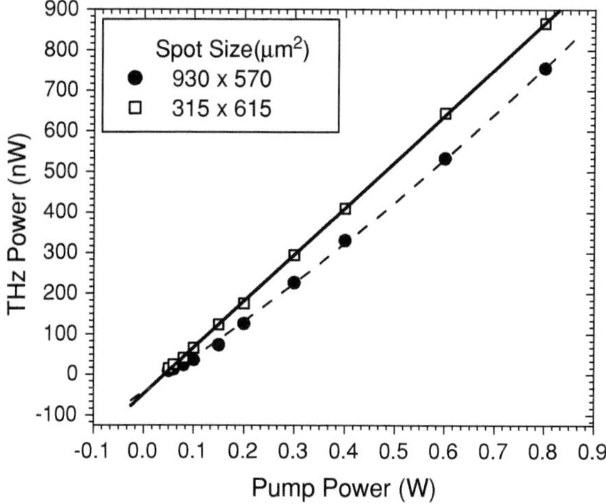

Fig. 5.6 Dependence of emitted THz power on pump power for $Ga_{0.44}In_{0.56}As$ crystal pumped with a maximum average power 800 mW. Experimental data for measurements at two different laser beam spot sizes are shown. The symbols are experimentally obtained data points and the solid and dotted lines are linear and polynomial fits of the data points respectively. The sample was photoexcited by a Yb:KYW laser operating at a center wavelength of 1.1 µm. The laser beam was incident on the sample at an angle of 30° with its surface-normal and laser spot had an elliptical shape

research [1]. However, as noted before in reference [1], many competing mechanisms play a role in the THz emission from $Ga_xIn_{1-x}As$ and it would be premature to conclude anything with certainty until further research is conducted in this regard.

5.3 Terahertz Emission from Semi-Large Aperture Fe-Doped $Ga_{0.69}In_{0.31}As$ Photoconducting Antenna Excited at 800 and 810 nm

In this section we present the experimental results of THz emission from semi-large aperture $Ga_{0.69}In_{0.31}As$ PC antennas excited by femtosecond Ti:S oscillator lasers. In the first part of this Sect. 5.3.1 we investigate the variation of THz time-domain signals emitted from $Ga_{0.69}In_{0.31}As$ PC antennas as a function of the applied bias field and incident laser power. These experiments are performed at an excitation wavelength of 800 nm and an average maximum laser power of 650 mW. In the second part of this Sect. 5.3.2 we measure the emitted THz power from $Ga_{0.69}In_{0.31}As$ PC antennas as a function of the applied bias field and the electrode spacing of the antenna. The experiments in this section are performed at excitation wavelengths of 800 and 810 nm. We conclude Sect. 5.3 by simulating the performance of a semi-large aperture $Ga_{0.69}In_{0.31}As$ PC antenna with 0.8 mm electrode spacing when it is photoexcited at 1.1 μm with a maximum average laser power of 1 W.

5.3.1 THz Time-Domain Measurements

Figure 5.7 shows the increase in peak amplitude of time-domain THz transients with increasing applied bias fields for a semi-large aperture $Ga_{0.69}In_{0.31}As$ PC antenna with 0.8 mm electrode gap spacing. The applied bias voltage ranged from 0 to 510 V and was increased in steps of 20 V. For the sake of clarity in Fig. 5.7, the THz transients for only four different values of the applied bias field are depicted. The data in Figs. 5.7, 5.8, and 5.9 were obtained using the THz time-domain spectroscopy system, the schematics of which are depicted in Fig. 4.1 in Chap. 4.

We note here that since the compact geometry of the experimental setup did not permit the insertion of a power meter close to the emitter surface, the laser powers shown in Figs. 5.7, 5.8, and 5.9 are measured at the output of the laser and do not reflect the power incident on the emitter surface. Nonetheless, Figs. 5.7, 5.8, and 5.9 provide valuable qualitative information about the dependence of the emitted THz signal on the applied bias field. In the later parts of this section we will quantitatively evaluate the THz power obtained from semi-large aperture $Ga_{0.69}In_{0.31}As$ PC antenna as a function of its physical parameters.

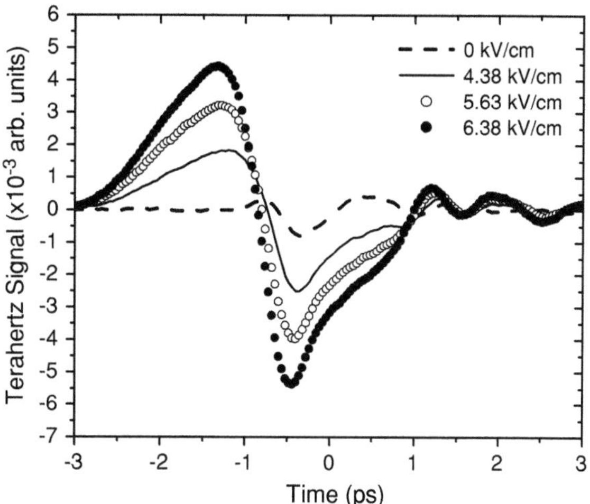

Fig. 5.7 Variation of emitted THz signal for different values of applied bias field for a Ga$_{0.69}$In$_{0.31}$As PC antenna photoexcited at 800 nm with an average laser output power of 500 mW. The electrode spacing of the antenna is 0.8 mm and the antenna is placed an angle of 45° with respect to the laser beam. The maximum applied bias field is 6.38 kV/cm

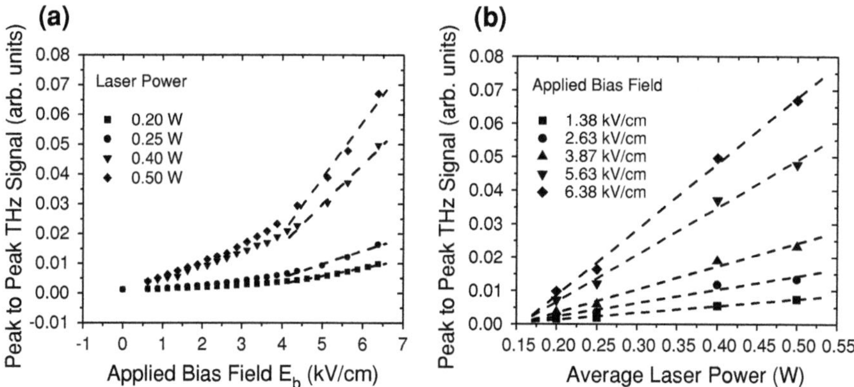

Fig. 5.8 Variation of peak-to-peak amplitude of THz signal with **a** applied bias field for different values of laser power, and **b** laser power for different values of applied bias voltage, for a Ga$_{0.69}$In$_{0.31}$As PC antenna with 0.8 mm electrode gap spacing. *Dashed lines* are linear fits of the data. The PC antenna was photoexcited at 800 nm with an average laser output power of 500 mW

Figures 5.8 and 5.9 depict the dependence of peak-to-peak amplitude of the emitted THz signal and photocurrent on applied bias voltage and incident laser power respectively, for a Ga$_{0.69}$In$_{0.31}$As PC antenna with 0.8 mm electrode gap spacing. In this context the peak-to-peak amplitude is defined as the sum of the amplitudes of the positive peak and the negative peak of the THz time-domain signals shown in Fig. 5.7. Figure 5.8a shows the dependence of the peak-to-peak amplitude of the emitted THz signal on the applied bias field, measured at a range of different laser powers.

The antenna was driven at bias field much lower than the bulk breakdown field ($\sim$12.5 kV/cm) to avoid electrical damage. From the plots in Fig. 5.8a, we

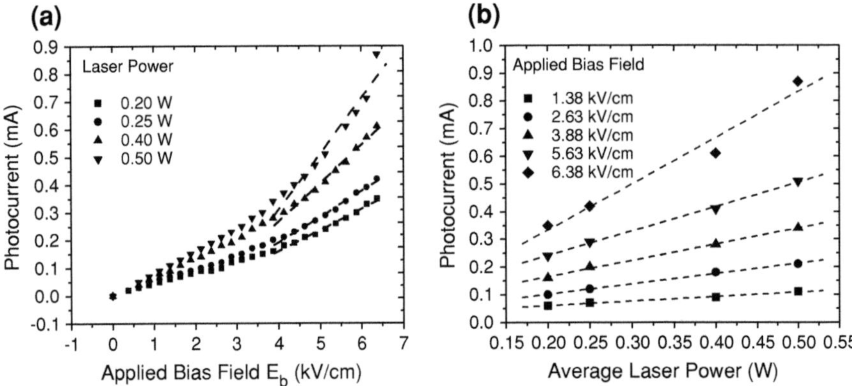

Fig. 5.9 Variation of photocurrent with **a** applied bias field for different values of laser power, and **b** laser power for different values of applied bias voltage, for a $Ga_{0.69}In_{0.31}As$ PC antenna with 0.8 mm electrode gap spacing. *Dashed lines* are linear fits of the data. The PC antenna was photoexcited at 800 nm with an average laser output power of 500 mW

observe that the variation of the peak-to-peak amplitude of the emitted THz signal is almost constant for bias fields less than 4 kV/cm and shows a steep linear rise beyond this point. This kind of threshold behavior can be attributed to the voltage drop at the electrodes of the PC antenna due to a non-ohmic resistance component at the metal–semiconductor interface. The linear dependence of the radiated THz signal on the applied bias field at higher bias field values, is expected from Eq. 2.40 in Sect. 2.2.1 (Chap. 2), which predicts that at low optical fluences (<0.5 μJ/ cm^2), the emitted THz signal varies linearly with the applied bias field for a given photoconductor material and at a given optical fluence. No saturation of the emitted THz radiation with applied bias voltage is observed for the present experimental conditions.

Figure 5.8b shows the dependence of the emitted THz signal on laser power for a range of applied bias fields. No saturation of the emitted THz radiation was observed for bias fields as high as 6.38 kV/cm. Saturation of the emitted THz signal from large or semi-large aperture PC antennas at high optical fluence (~ 10 μJ/cm^2) has been observed experimentally before [26, 27], and effectively limits the THz power output from such a device. The saturation of the emitted THz signal proceeds either by the screening of the applied bias field by the generated THz radiation or by the screening of bias field by the space charge induced by separating carriers. The space charge screening effect is seen to have diminishing influence on the THz emission from large aperture (>1 mm) PC antennas [28, 29]. This is because at larger electrode spacing the transit time across the antenna gap is much longer (~ 10 ps in a 1 mm gap GaAs PC antenna emitter), and the emitter sees a screened bias field only after the THz radiation has emitted. Thus, the space-charge screening effect in large or semi-large aperture PC antennas is only dominant at very high values of optical fluence, typically of the order of few

$\mu J/cm^2$. On the other hand, the saturation fluence for the terahertz screening of the applied bias field to take effect, has been estimated to be 9.6 $\mu J/cm2$ in a large aperture GaAs PC antenna emitter with 1.9 mm electrode separation [3]. For the present experimental purpose, the optical fluence incident on the emitter for the highest laser power output is 0.3 $\mu J/cm^2$. This explains why we do not observe the emitted THz radiation to saturate with increasing laser power in Fig. 5.8b.

The relation between photocurrent and applied bias voltage was measured at a range of laser powers as shown in Fig. 5.9a. The dependence of photocurrent on bias voltage shows similar threshold behavior observed in the dependence of emitted THz radiation on bias voltage. This is expected from Eqs. 2.37 and 2.38 in Sect. 2.2.1 (Chap. 2) which predict that the peak to peak amplitude of the emitted THz radiation shows similar dependence on bias field as the photocurrent. At higher laser powers, the photocurrent rises more steeply due to substrate heating and consequent increase in conductivity.

Figure 5.9b shows the dependence of photocurrent on average laser power for a range of applied bias voltages. The linear dependence of photocurrent on laser power confirms that at low optical fluences, such as those employed in our experiments, Eq. 2.19a can be approximated to Eq. 2.41 $(\mathbf{J}_s(t) \cong \sigma_s(t)\mathbf{E}_b)$. It further confirms that the conductivity σ_s varies linearly with the optical fluence F_{opt}. as depicted in Eq. 2.39 $\left(\sigma_s \max = \frac{e(1-R)\mu \cos\theta_1 F_{\mathrm{opt}}}{\hbar\omega}\right)$.

The slopes of the plots increase with increasing bias field since the photocurrent is proportional to the applied bias as per Eq. 2.41 $(\mathbf{J}_s(t) \cong \sigma_s(t)\mathbf{E}_b)$.

5.3.2 THz Power Measurements

The THz power measurement data presented in this section was obtained using the setup whose schematics are presented in Fig. 4.2 in Chap. 4. Figure 5.10 shows the variation of emitted THz power as a function of the square of the applied bias field for different electrode gap spacing. The PC antennas are excited at 800 nm wavelength and the average laser power incident on the photoconducting antenna is 250 mW. We observe that the emitted THz power increases linearly with the square of the applied bias voltage as predicted by Eq. 2.47 in Chap. 2. We also observe from Fig. 5.10 that the emitted THz power increases with the electrode gap spacing of the PC antenna.

Figure 5.11 plots the variation of the emitted THz power as a function of the aperture area of the PC antenna. Here, we define the aperture area as a circular area with diameter equal to the electrode gap spacing of the PC antenna. The data in Fig. 5.11 was obtained under an applied bias field of 8.75 kV/cm and an excitation laser power of 250 mW. We observe that the emitted THz power scales linearly with the aperture area of PC antenna. This result in accordance with Eq. 2.49 which predicts that for a given photoconductor, if the applied bias field $\mathbf{E}_b$. and the incident laser power $\mathbf{P}_{\mathrm{opt}}$ are kept constant, the average radiated THz power scales

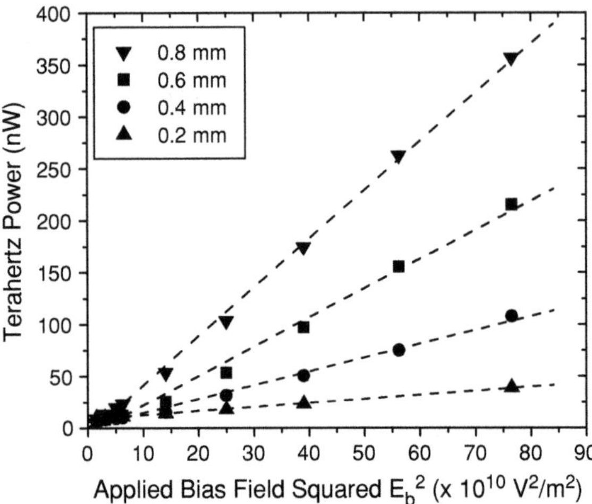

Fig. 5.10 Variation of emitted THz power with the square of the applied bias field in a $Ga_{0.69}In_{0.31}As$ PC antenna for different electrode gap spacing. The PC antennas were photoexcited at 800 nm wavelength and at an average laser power of 250 mW. *Dashed lines* are linear fits of the data

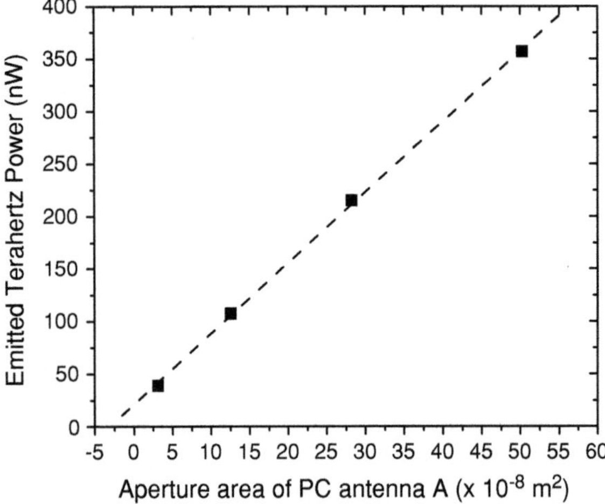

Fig. 5.11 Variation of emitted THz power with aperture area of $Ga_{0.69}In_{0.31}As$ PC antenna with an applied bias field of 8.75 kV/cm. The PC antennas were photoexcited at 800 nm wavelength and at an average laser power of 250 mW. *Dashed lines* are linear fits of the data [21]

linearly as the antenna gap spacing A. The dashed lines in Figs. 5.10 and 5.11 are the linear fits of the data. The slopes of the linear fits of the data in Fig. 5.10 have been presented in Table 5.4.

From the fitting parameters in Table 5.4 and the known physical quantities in Eq. 2.47 it is possible to estimate the value of the ultrafast mobility of Fe-doped $Ga_{0.69}In_{0.31}As$. Table 5.5 lists the physical quantities in Eq. 2.47 along with their numerical values and dimensions in SI units. The estimated values of ultrafast carrier mobility in $Ga_{0.69}In_{0.31}As$ calculated from Eq. 2.47 and Tables 5.4 and 5.5 are listed in Table 5.6.

We note from Table 5.6 that the estimated values of ultrafast carrier mobility in $Ga_{0.69}In_{0.31}As$ are about two orders of magnitude higher than the dc mobility of the

Table 5.4 Fitting parameters for data in Fig. 5.10

Electrode gap spacing (mm)	0.8	0.6	0.4	0.2
Slope (W/V^2)	$4.69 \times 10^{-19} \pm$ 7.48×10^{-21}	$2.82 \times 10^{-19} \pm$ 9.80×10^{-21}	$1.33 \times 10^{-19} \pm$ 4.75×10^{-21}	$3.89 \times 10^{-20} \pm$ 6.61×10^{-22}

Table 5.5 List of physical parameters in Eq. 2.47

Description and symbol of physical quantities	SI units	Value
Electron charge (e)	Coulomb	1.6×10^{-19}
Infrared refractive index of $Ga_{0.69}In_{0.31}As$ [25] (n)	–	3.39
Laser power $\left(P_{opt}\right)$	Watt	0.250
Laser Photon Energy at 800 nm ($\hbar\omega$)	Joule (Watt-Second)	2.48×10^{-19}
Carrier lifetime (τ_c)[a]	Second	0.306×10^{-12}
Time period between consecutive laser pulses ($Trep$)	Second	1.2×10^{-8}
Distance between THz emitter and detector (z)	Meter	0.1860
Laser focal spot diameter	Meter	1×10^{-3}
Permittivity of free space (ϵ_0)	$(Coulomb)^2/(Newton\text{-}m^2)$	8.85×10^{-12}
Velocity of light in vacuum (c)	Meter/second	3×10^8

[a] Carrier lifetime was obtained from transient photo-reflectivity measurements explained in Sect. 5.1 of this chapter

Table 5.6 Estimated values of ultrafast carrier mobilities in $Ga_{0.69}In_{0.31}As$

Electrode gap spacing (mm)	0.8	0.6	0.4	0.2
Mobility μ_t (m^2/V-s)	106.8 ± 0.9	110.5 ± 1.9	113.8 ± 2.0	123.1 ± 1.0

material (2395 cm^2/Vs). The mobility referred to as the "ultrafast mobility" μ_t in this experimental work only takes into account those scattering processes that occur on a timescale relevant to the THz emission process. We define the ultrafast mobility as the time-averaged value of the time-dependent carrier mobility over a time period τ_{THz} relevant to the THz emission process.

$$\mu_t = \frac{1}{\tau_{THz}} \int_0^{\tau_{THz}} \mu(t)dt \tag{5.1}$$

Here τ_{THz} is the time period of the emitted THz pulse. A higher value of ultrafast mobility compared with the dc Hall mobility was also observed earlier in femtosecond optical pump-probe experiments in n-GaAs [30]. This was attributed to the different scattering mechanisms in the dc Hall regime and the ultrafast regime. The main scattering mechanism of non-equilibrium electron gas is governed by electron–hole scattering [30], while that of an electron gas in equilibrium drifting under a weak and static electric field (such as that found in Hall measurements) is governed by impurity and polar phonon scattering as well as carrier-carrier scattering. The ultrafast mobility is higher than the dc mobility since not all of the scattering rates measured in the dc mobility are accounted for in the ultrafast mobility [31].

In Table 5.6 we also observe that the ultrafast mobility increases with the decrease in electrode gap spacing in Fe-doped $Ga_{0.69}In_{0.31}As$. If the laser beam diameter at the emitter is kept constant (such as in our experiments), a decrease in electrode spacing implies a decrease in photogenerated carrier density, since a smaller active area is exposed to the laser beam for photoexcitation. For lower photocarrier density, carrier-carrier scattering decreases and this results in the observed increase in ultrafast mobility in Fe-doped $Ga_{0.69}In_{0.31}As$ PC antenna as the electrode gap spacing is reduced.

Previously, the ultrafast mobility in GaAs [32] and $Ga_{0.47}In_{0.53}As$ [33] was estimated by fitting the transmission data obtained using optical pump THz probe (OPTP) technique to the Drude model. A substantially reduced Drude-fit mobility compared to the Hall mobility was observed in either case, and was attributed to the additional momentum relaxation associated with electron–hole scattering events due to the high photocarrier densities ($\sim 10^{18}/cm^3$) under which the OPTP experiments were performed. A similar argument was also presented by Heyman et al. [34] to explain the observed discrepancy between the dc Hall mobility and the ultrafast mobility estimated using the THz photo-Hall measurements.

Since carrier mobility is associated with carrier-impurity scattering processes, it is expected that materials with good crystalline quality exhibit high carrier mobility. In a theoretical study of carrier dynamics in low temperature grown (LT) GaAs using Monte Carlo simulations by Reklaitis et al. [35] it was shown that the high carrier mobility in LT GaAs could be attributed to the perfect crystallinity and the sub-picosecond carrier trapping times in this material. In their work [35], Reklaitis et al. demonstrated that at low electric fields (less than ~ 7.5 kV/cm) the drift velocity (and consequently the mobility) is larger due to reduction of ionized impurity scattering of the photocarriers created high into the conduction band from where they cannot cool down to thermal energies during their short lifetimes. At high electric fields (~ 10 kV/cm) the drift velocity (and hence the mobility) of the carriers can attain values several times larger than the steady state drift velocity due to the velocity overshoot effect. Since carrier lifetime in LT GaAs is of the order of or less than or 1 ps, the electron energy never reaches its stationary value at a given electric field, and the drift velocity remains overshot as long as the photocarriers are present in the material. Thus, the enhanced carrier mobility in LT GaAs is caused by the ultrafast trapping of electrons by deep traps in the material that efficiently reduce some of the scattering mechanisms. A similar argument can be put forward to explain the high estimated values of ultrafast mobility in the Fe-doped $Ga_{0.69}In_{0.31}As$ crystals used in our experiments which also exhibit sub-picosecond carrier lifetimes. The Fe-doped $Ga_{0.69}In_{0.31}As$ bulk crystal employed in our work possesses high crystalline perfection due to near-equilibrium

Table 5.7 Comparison of estimated ultrafast carrier mobilities in GaAs and $Ga_{0.69}In_{0.31}As$

Material	Experimental technique	Excitation wavelength (nm)	Photocarrier concentration (/cm^3)	μ_t (cm^2/Vs)	μ_{Hall} (cm^2/Vs)
GaAs [34]	THz photo Hall measurement	800	$\sim 10^{17}$	4400 ± 600	6950
GaAs [32]	OPTP measurement	660	0.4×10^{18}	1300	8000
GaAs [26]	THz emission measurement	618	–	220	8500
GaAs [36]	Optical pump-probe measurement	815	1.0×10^{18}	3500	2300
$Ga_{0.47}In_{0.53}As$ [33]	OPTP measurement	810	2.5×10^{15}	5000 ± 500-6000 ± 500	10000
Material developed for present research					
$Ga_{0.69}In_{0.31}As$	THz emission measurement	800	$\sim 10^{17}$	$\sim 10^6$	2395

growth process whereas MBE grown epilayers of $Ga_xIn_{1-x}As$ on binary substrates may suffer from poor crystalline quality due to misfit dislocations that originate at the growth interface and propagate into the device layers as a result of the lattice mismatch between the epilayer and the substrate. Typical misfit dislocation density is in the range of 10^6–10^9 cm^{-2}.

A comparison between estimated ultrafast carrier mobilities and dc Hall-mobilities for GaAs and $Ga_xIn_{1-x}As$ involving different experimental procedures is presented in Table 5.7. The high values of ultrafast mobility in Fe-doped $Ga_{0.69}In_{0.31}As$ estimated from our experiments establish this material as a strong candidate for THz PC antenna applications.

Next, we compare the performance of Fe doped $Ga_{0.69}In_{0.31}As$ PC antenna with that of semi-insulating (SI) GaAs PC antenna and bare InAs crystal in terms of emitted THz power. The electrode spacing for the PC antennas is 0.4 mm. SI GaAs and InAs are commonly used THz emitter materials in THz time-domain spectroscopy. Figure 5.12 shows the variation of emitted THz power with applied bias field for each of these materials. The InAs sample was placed at an angle of $\sim 45°$ with respect to the incident laser beam. The $Ga_{0.69}In_{0.31}As$ and SI GaAs PC antennas were photoexcited by the laser beam at normal incidence. The average power of the incident laser beam was 250 mW. The results of the measurement are summarized in Table 5.8.

For a given value of the incident laser power, the THz output power from SI GaAs and $Ga_{0.69}In_{0.31}As$ PC antennas is limited by the breakdown fields in air associated with these materials at which surface flashover occurs. For GaAs and $Ga_{0.69}In_{0.31}As$ PC antennas, this value is estimated to 12.5 and 12 kV/cm [37] respectively. However, under the present experimental conditions, it was possible to apply a maximum bias field of only 5 kV/cm to the SI GaAs PC antenna. Beyond this point, the current increases above the current limit (~ 10 mA) of the

Table 5.8 Comparison of emitted THz power between SI GaAs, $Ga_{0.69}In_{0.31}As$, and InAs from Fig. 5.11

Laser Power (W)	Bias Field (kV/cm)	Material	THz Power (nW)
0.25	5	SI GaAs	106.4
	0	InAs	13.7
	8.75	$Ga_{0.69}In_{0.31}As$	107.5

high voltage power supply and the power supply shuts down. This nonlinear increase in photocurrent at higher applied bias voltages (>4 kV) in SI GaAs sets in as a result of persistent photoconductivity and substrate heating. The persistent photoconductivity phenomena in SI GaAs arises because the photo generated carriers do not decay completely between consecutive laser pulses as a result of long carrier recombination lifetime (~ 100 ns) compared to the time period of the laser pulse (~ 12 ns). The higher conductivity in SI GaAs also accounts for the steep rise of the emitted THz power from this material at lower bias voltages (1.5–5 kV/cm) compared to that from $Ga_{0.69}In_{0.31}As$.

In $Ga_{0.69}In_{0.31}As$ PC antenna, the photocurrent remains much lower due to the shorter recombination lifetime (~ 0.3 ps), thus reducing the ohmic heating and increasing the threshold voltage for thermal runaway. For this reason, it is possible to apply a higher bias voltage to $Ga_{0.69}In_{0.31}As$ than to SI GaAs. The emitted THz power obtained from $Ga_{0.69}In_{0.31}As$ and SI GaAs PC antennas are comparable for the highest applied bias fields in each case (Table 5.8). However in case of $Ga_{0.69}In_{0.31}As$, it is possible to operate the PC antenna at higher bias fields, close to the bulk breakdown field for the material (12.5 kV/cm) and thus extract higher THz emission power.

Compared to $Ga_{0.69}In_{0.31}As$ and SI GaAs PC antennas, the InAs sample has a much lower value of emitted THz power. Indeed, it is possible to enhance the THz emission from InAs by application of an external magnetic field. However, as shown by Reid et al. [3] in a comparative study of THz emission from SI GaAs PC antenna and InAs, the efficiency of this process is low. From Fig. 5.12 and Table 5.8 we are thus able to demonstrate that $Ga_{0.69}In_{0.31}As$ PC antenna has a superior performance compared to conventional THz emitters such as SI GaAs PC antenna and InAs surface emitter.

It is possible to further enhance the THz emission from PC antennas by attaching a high resistivity silicon lens on the backside of the antenna (as shown in Fig. 4.3b) for efficient collimation of the THz beam from the PC antenna emitter to the detector. Attaching a silicon lens to the PC antenna emitter also reduces the Fresnel reflection losses at the semiconductor-air interface and allows for efficient coupling of the THz radiation onto free space. For our experiments we attached a 6 mm diameter, hemispherical Si lens on the backside of the 0.8 mm $Ga_{0.69}In_{0.31}As$ PC antenna with a thin layer of vacuum grease.

Since THz radiation is strongly absorbed by water [38], further enhancement of emitted THz power can be effected by enclosing the THz emitter, detector, and collimating optics inside a nitrogen-purged chamber. The downside is that, a

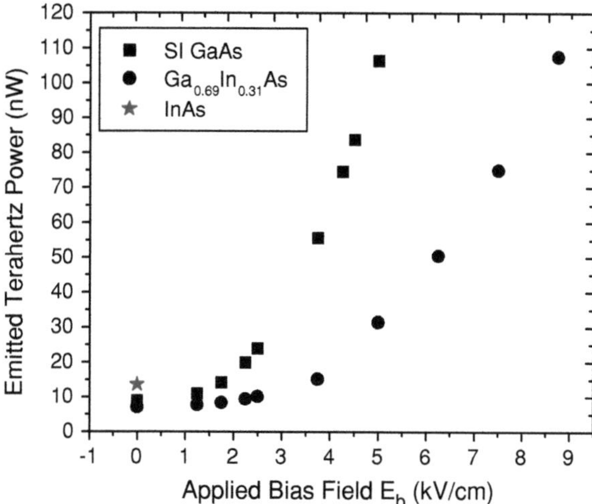

Fig. 5.12 Variation of emitted THz power with applied bias field for InAs, and $Ga_{0.69}In_{0.31}As$ and SI GaAs PC antennas with 0.4 mm electrode spacing. The incident laser power is 250 mW. The emitters are photoexcited at 800 nm. The breakdown fields in air, at which surface flashover occurs in SI GaAs and $Ga_{0.69}In_{0.31}As$ are 12 and 12.5 kV/cm respectively. Bias fields higher than 5 kV/cm cannot be applied in SI GaAs due to thermal runaway [21]

Golay cell detector, such as the one used in our experimental setup is very sensitive to mechanical vibrations, and the turbulence caused by the nitrogen flow inside the chamber may introduce unwanted noise into the detected signal. Nonetheless, it is possible to estimate the effects of nitrogen purging on the emitted THz power from the enhancement factor of the THz time-domain signal obtained under nitrogen-purged conditions from a typical time-domain setup (Fig. 4.1). The THz transient obtained from such a setup in air and under nitrogen-purged conditions from InAs is shown in Fig. 5.13.

By comparing the peak-to-peak amplitudes of the THz transients obtained in air and under nitrogen-purged conditions we can compute the enhancement factor of the emitted THz signal under nitrogen-purged condition. From ten such instances we computed the average enhancement factor to be 1.61 ± 0.26. Since $P \propto |E|^2$, an enhancement of factor of 1.61 ± 0.26 in emitted peak THz electric field implies an enhancement of $\sim 2.59 \pm 0.84$ in emitted peak THz power. In the following measurement performed on $Ga_{0.69}In_{0.31}As$ PC antenna with 0.8 mm electrode spacing, we multiply all data points with this scaling factor to simulate the conditions of nitrogen purging. Figure 5.14 shows the variation of emitted THz power as a function of the square of applied bias field, at a range of laser powers, for a $Ga_{0.69}In_{0.31}As$ PC antenna with 0.8 mm electrode spacing. We used a Ti:S oscillator laser ($\lambda = 810$ nm) with higher average output power to photoexcite the PC antenna. A linear fit is obtained for the data shown in Fig. 5.14 and the slopes are listed in Table 5.9.

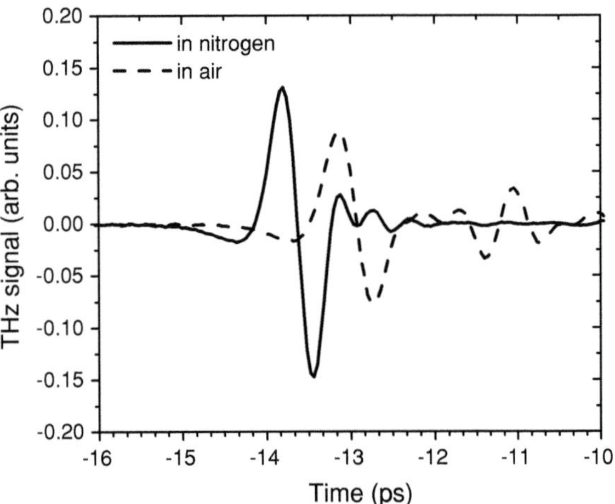

Fig. 5.13 THz time-domain signals from InAs in air (*dashed line*) and under nitrogen purged conditions (*solid line*). The sample was photoexcited at 800 nm

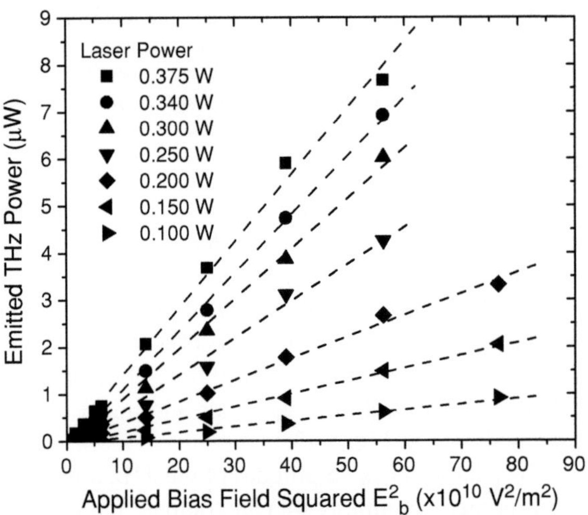

Fig. 5.14 Variation of emitted THz power with the square of applied bias field, for a range of laser powers for a $Ga_{0.69}In_{0.31}As$ PC antenna with 0.8 mm electrode spacing. *Dashed lines* are linear fits of the data. The PC antenna is excited at a wavelength of 810 nm and at a maximum average laser power of 375 mW. The data is multiplied by the number 2.59 (see text for explanation) to simulate nitrogen-purged conditions. A silicon lens was attached on the backside of the antenna to facilitate enhanced collimation of the THz beam from the PC antenna to the detector

At high incident laser powers (>200 mW) and high bias field values (>7.5 kV/cm) the photocurrent in the antenna increases nonlinearly due to ohmic heating of the semiconductor substrate and thermal runaway sets in. This limits the bias field that can be applied to the PC antenna at such high incident laser powers and

Table 5.9 Slope of the linear fit of data shown in Fig. 5.13

Laser power (W)	Slope (W-m^2/V^2)
0.375	$1.42 \times 10^{-17} \pm 3.64 \times 10^{-19}$
0.340	$1.24 \times 10^{-17} \pm 1.97 \times 10^{-19}$
0.300	$1.07 \times 10^{-17} \pm 2.74 \times 10^{-19}$
0.250	$7.83 \times 10^{-18} \pm 2.73 \times 10^{-19}$
0.200	$4.57 \times 10^{-18} \pm 1.06 \times 10^{-19}$
0.150	$2.71 \times 10^{-18} \pm 8.05 \times 10^{-20}$
0.100	$1.18 \times 10^{-18} \pm 5.26 \times 10^{-20}$

consequently limits the THz output power from the antenna. Substrate heating effects and thermal runaway can be reduced by resorting to an active cooling method such as a Peltier cooling mechanism.

Since the emitted THz power scales linearly with the square of the applied bias field, it is possible to extract higher THz powers (than those depicted in Fig. 5.14) from the Ga$_{0.69}$In$_{0.31}$As PC antenna with 0.8 mm electrode spacing, by operating the antenna close to its breakdown field (12.5 kV/cm). However, it was not possible to conduct the experiment under the current conditions since device heating made it impossible to apply bias fields higher than 7.5 kV/cm at incident laser powers higher than 200 mW. Nonetheless, it is possible to simulate the performance of the Ga$_{0.69}$In$_{0.31}$As PC antenna with 0.8 mm electrode spacing at a high bias field of 11.5 kV/cm for different values of incident laser powers. This can be done by extrapolating each of the linear fits in Fig. 5.14 (from the slopes of the fits in Table 5.9) to calculate the THz power obtained from a Ga$_{0.69}$In$_{0.31}$As PC antenna with 0.8 mm electrode spacing, at an applied bias field of 11.5 kV/cm. These calculated data points have been plotted as a function of the incident laser power in Fig. 5.15.

We observe, that the simulated THz power from the Ga$_{0.69}$In$_{0.31}$As PC antenna with 0.8 mm electrode spacing, driven at a bias field of 11.5 kV/cm, varies sublinearly with the square of the laser power as we go to higher ($>$340 mW) incident laser powers. The deviation from linearity has been shown by fitting a straight line to the first five data points. For a given bias field and electrode gap spacing, a linear variation of the THz power with the square of the laser power is expected from Eq. 2.49 $\left(\text{P} \propto P_{opt}^2 \right)$ in Chap. 2. However at a high applied bias field and high incident laser powers (such as those employed in the present experiment), a sublinear dependence is possible due to screening of the applied bias field by the generated THz radiation or by the by the space charge induced by separating carriers. Such behavior has also been observed before in large aperture GaAs and InP based PC antennas [26].

In order to predict the performance of the Ga$_{0.69}$In$_{0.31}$As PC antenna at higher incident laser powers such as those obtained from Yb:KYW lasers (Table 5.11), we fit data in Fig. 5.14 to an equation of the form, $y = P_1 x/(P_2 + x)$ (shown by the dashed curve). Here x and y denote the quantities depicted (namely the emitted THz power and the square of the laser power) on the x and y axes of the plot in Fig. 5.15. The fitting parameters P_1 and P_2 are listed in Table 5.10. Employing the

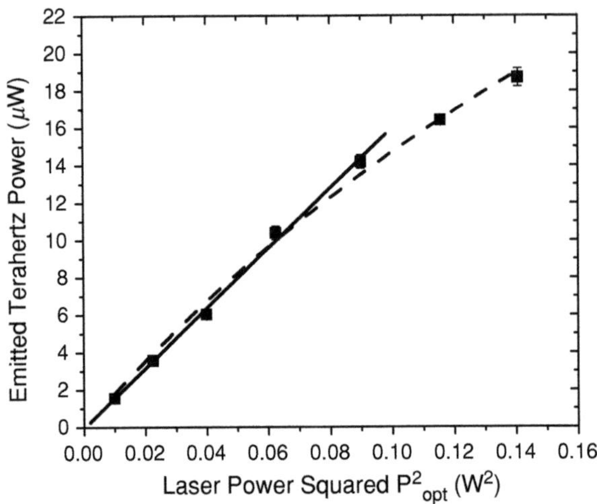

Fig. 5.15 Variation of simulated THz power as a function of the square of incident laser power, for a $Ga_{0.69}In_{0.31}As$ PC antenna with 0.8 mm electrode spacing, and a bias field of 11.5 kV/cm. The data points are obtained by extrapolating the linear fits in Fig. 5.14 for each value of incident laser power. *Dashed line* indicates a polynomial fit of the data. *Solid line* is a linear fit of the first five data points. The *solid line* is drawn to indicate that the emitted THz power deviates from its expected linear dependence on the square of the laser power

Table 5.10 Fitting parameter for polynomial fit of data shown in Fig. 5.5	Parameter	Value
	P_1 (W)	$7.0 \times 10^{-5} \pm 2.0 \times 10^{-5}$
	P_2 (W^2)	0.37209 ± 0.11469

Table 5.11 Physical parameters for Yb:KYW laser [39]	Center wavelength (μm)	Power (W)	Pulse duration (fs)	Repetition rate (MHz)
	1.1	2	100–150	85

fitting parameters in Table 5.10 and the Yb:KYW laser parameters in Table 5.11 we estimate that it is possible to extract an output THz power of 51 μW from a $Ga_{0.69}In_{0.31}As$ PC antenna with 0.8 mm electrode spacing, biased at 11.5 kV/cm, at an incident laser power of 1 W.

This result is indeed promising since our $Ga_{0.69}In_{0.31}As$ PC antenna has an estimated THz output power that is within a factor of 10 from realizing a strong THz source with half a millwatt of output power. The estimated output power is also about three orders of magnitude higher (Table 5.8) than that obtained from InAs surface emitter which is commonly used in THz time-domain spectroscopy systems. Our experimental results further indicate that Fe-doped $Ga_{0.69}In_{0.31}As$ exhibits high ultrafast mobility and sub-picosecond carrier lifetimes that are much desired in THz PC antenna emitters alongside with a high dark resistivity. Together, these results successfully establish $Ga_{0.69}In_{0.31}As$ PC antenna as a promising candidate for high output power, MHz repetition rate THz emitter driven by femtosecond lasers.

References

1. Ko, Y., Sengupta, S., Tomasulo, S., Dutta, P., Wilke, I.: Emission of terahertz-frequency electromagnetic radiation from bulk $Ga_xIn_{1-x}As$ crystals. Phys. Rev. B **78**, 035201 (2008)
2. Ko, Y.: Electrical, optical, and THz emission studies of GaxIn1-xAs bulk crystals. Department of Electrical, Computer, and Systems Engineering, Doctoral Thesis, Rensselaer Polytechnic Institute, Troy, NY (2007)
3. Reid, M., Fedosejevs, R.: Quantitative comparison of terahertz emission from (100) InAs surfaces and a GaAs large-aperture photoconductive switch at high fluences. Appl. Opt. **44**, 149 (2005)
4. Korn, T., Franke-Wiekhorst, A., SchnÄull, S., Wilke, I.: Characterization of nanometer As-clusters in low-temperature grown GaAs by transient reflectivity measurements. J. Appl. Phys. **91**, 2333 (2002)
5. Cho, G.C., Kütt, W., Kurz, H.: Subpicosecond time-resolved coherent-phonon oscillations in GaAs. Phys. Rev. Lett. **65**, 764 (1990)
6. Gupta, S., Frankel, M.Y., Valdmanis, J.A., Whitaker, J.F., Smith, F.W., Calawa, R.: Subpicosecond carrier lifetime in GaAs grown by molecular beam epitaxy at low temperatures. Appl. Phys. Lett. **59**, 3276 (1991)
7. Ganikhanov, F., Lin, G.-R., Chen, W.-C., Chang, C.-S., Pan, C.-L.: Subpicosecond carrier lifetimes in arsenic-ion-implanted GaAs. Appl. Phys. Lett. **67**, 3465 (1995)
8. Giniunas, L., Danielius, R., Tan, H.H., Jagadish, C., Adomavicius, R., Krotkus, A.: Electron and trap dynamics in As-ion-implanted and annealed GaAs. Appl. Phys. Lett. **78**, 1667 (2001)
9. Prabhu, S.S., Vengurlekar, A.S.: Dynamics of the pump-probe reflectivity spectra in GaAs and GaN. J Appl. Phys. **95**, 7803 (2004)
10. Das Sarma, S., Jain, J.K., Jalabert, R.: Theory of hot-electron energy loss in polar semiconductors: role of plasmon-phonon coupling. Phys. Rev. B **37**, 6290 (1988)
11. Bair, J.E., Cohen, D., Krusius, J.P., Pollock, C.R.: Femtosecond relaxation of carriers generated by near-bandgap optical excitation in compound semiconductors. Phys. Rev. B **50**, 4355 (1994)
12. Shah, J.: Hot electrons and phonons under high intensity photoexcitation of semiconductors. Solid State Electron. **21**, 43 (1978)
13. Sucha, G., Bolton, S.R., Chemla, D.S., Sivco, D.L., Cho, A.Y.: Carrier relaxation in InGaAs heterostructures. Appl. Phys. Lett. **65**, 1486 (1994)
14. Ilczuk, E., Korona, K.P., Babinski, A., Kuhl, J.: Dynamics of photoexcited carriers in GaInAs/GaAs quantum dots. Acta Phys. Polonica A **100**, 379 (2001)
15. Shah, J., Leheny, R.F., Nahory, R.E., Pollack, M.A.: Hot carrier relaxation in photoexcited $In_{0.53}Ga_{0.47}As$. Appl. Phys. Lett. **37**, 475 (1980)
16. Kash, K., Shah, J.: Carrier energy relaxation in In0.53Ga0.47As determined from picosecond luminiscence studies. Appl. Phys. Lett. **45**, 401 (1984)
17. Hamanaka, Y., Nishiwaki, D., Nonogaki, Y., Fujiwara, Y., Takeda, Y., Nakamura, A.: Femtosecond relaxation dynamics of hot carriers in photoexcited $In_{0.53}Ga_{0.47}As$. Phys. B **272**, 391 (1999)
18. Nishiwaki, D., Hamanaka, Y., Nonogaki, Y., Fujiwara, Y., Takeda, Y., Nakamura, A.: Hot carrier relaxation dynamics in $In_{0.53}Ga_{0.47}As$ studied by femtosecond pump-probe spectroscopy. J. Lumin. **83–84**, 49 (1999)

19. Jo, S.J., Ihn, S.G., Song, J.I., Yee, K.J., Lee, D.H.: Carrier dynamics of low-temperature-grown $In_{0.53}Ga_{0.47}As$ on GaAs using an InGaAlAs metamorphic buffer. Appl. Phys. Lett. **86**, 1–111903 (2005)
20. Lobentanzer, H., Ruhle, W.W., Stolz, W., Ploog, K.: Hot carrier phonon interaction in three and two dimensional $Ga_{0.47}In_{0.53}As$. Solid State Commun. **62**, 53 (1987)
21. Reprinted with permission from Sengupta, S., Wilke, I., Dutta, P. S.: Femtosecond carrier dynamics in native and high resistivity iron-doped $Ga_xIn_{1-x}As$. J. Appl. Phys. **107**, 033104 (2010). Copyright 2010, American Institute of Physics
22. Gupta, S., Whitaker, J.F.: Ultrafast carrier dynamics in III-V semiconductors grown by molecular-beam epitaxy at very low substrate temperatures. IEEE J. Quant. Electron. **28**, 2464 (1992)
23. Ascazubi, R., Wilke, I., Cho, S., Lu, H., Schaff, W.J.: Ultrafast recombination in Si-doped InN. Appl. Phys. Lett. **88**, 1–112111 (2006)
24. Chen, F., Cartwright, A., Lu, H., Schaff, W.: Ultrafast carrier dynamics in InN epilayers. J. Crystal Growth **269**, 10 (2004)
25. Goldberg, Y.A., Shmidt, N.M.: Gallium indium arsenide. In: Levinshtein, M., Rumayantsev, S., Shur, M.S. (eds.) Handbook Series on Semiconductor Parameters, vol. 2, p. 62. World Scientific, Singapore (1999)
26. Darrow, J.T., Zhang, X.-C., Auston, D.H., Morse, J.D.: Saturation properties of large aperture photoconducting antennas. IEEE J. Quant. Electron. **28**, 1607 (1992)
27. Benicewicz, P.K., Roberts, J.P., Taylor, A.J.: Scaling of terahertz radiation from large-aperture biased photoconductors. J. Opt. Soc. Am. B **11**, 2533 (1994)
28. Rodriguez, G., Taylor, A.J.: Screening of the bias field terahertz generation from photoconductors. Opt. Lett. **21**, 1046 (1996)
29. Taylor, A.J., Rodriguez, G., Some, D.: Ultrafast carrier dynamics in large-aperture photoconductors. Opt. Lett. **22**, 715 (1997)
30. Hase, M.: Carrier mobility in polar semiconductor measured by an optical pump-probe technique. Appl. Phys. Lett. **94**, 1–112111 (2009)
31. Ascazubi, R.: THz emission spectroscopy of narrow bandgap semiconductors. Department of Physics, Applied Physics, and Astronomy, Doctoral Thesis, Rensselaer Polytechnic Institute, Troy, NY (2005)
32. Greene, B.I., Saeta, P.N., Dykaar, D.R., Schmitt-Rink, S., Chuang, S.L.: Far-infrared light generation at semiconductor surfaces and its spectroscopic applications. IEEE J. Quant. Electron. **28**, 2302 (1992)
33. Ralph, S.E., Chen, Y., Woodall, J., McInturff, D.: Subpicosecond photoconductivity in $In_{0.53}Ga_{0.47}As$: intervalley scattering rates observed via THz spectroscopy. Phys. Rev. B. **54**, 5568 (1996)
34. Heyman, J.N., Bell, D., Khumalo, T.: Terahertz photo-Hall measurements of carrier mobility in GaAs and InP. Appl. Phys. Lett. **88**, 1–162104 (2006)
35. Reklaitis, A., Krotkus, A., Grigaliūnaitė, G.: Enhanced drift velocity of photoelectrons in a semiconductor with ultrafast carrier recombination. Semiconduct. Sci. Technol. **14**, 945 (1999)
36. Hase, M., Nakashima, S., Mizoguchi, K., Harima, H., Sakai, K.: Ultrafast decay of coherent plasmon-phonon coupled modes in highly doped GaAs. Phys. Rev. B **60**, 16526 (1999)
37. Wei, S., Zhenxian, L., Jun, F., Chuanxiang, X.: Insulation protection of high-power GaAs photoconductive switch. In: Proceedings of the 5th International Conference on Properties and Applications of Dielectric Materials, May 25–30, Seoul, Korea (1997)
38. van-Exter, M., Fattinger, C., Grischkowsky, D.: Terahertz time-domain spectroscopy of water vapor. Opt. Lett. **14**, 1128 (1989)
39. Data obtained from *QPeak Inc*

Chapter 6
Conclusions and Future Outlook

This dissertation contains results obtained from investigating pulsed THz emission characteristics of semi-large aperture photoconducting (PC) antennas fabricated on Fe-doped bulk $Ga_{0.69}In_{0.31}As$ substrate. The research is aimed at evaluating the impact of physical properties of a semi-large aperture $Ga_{0.69}In_{0.31}As$ PC antenna upon its THz generation efficiency, and is motivated by the ultimate goal of developing a pulsed THz radiation source with MHz repetition rates, capable of operating at wavelengths between 1 and 1.5 μm, and having an average output power of 1 mW.

To this end we investigate THz emission characteristics of as-grown $Ga_xIn_{1-x}As$ surface emitters photoexcited at 1.1 μm. Our experimental results demonstrate that an average output THz power of 867 nW can be extracted from a $Ga_{0.44}In_{0.56}As$ surface emitter when it is photoexcited with 800 mW of laser power. This motivated us to fabricate semi-large aperture PC antennas on GaInAs since it has been shown in prior research [1] that these types of THz emitters have a higher efficiency for converting the photoexcitation laser power to output THz power. A primary challenge in fabricating semi-large aperture PC antennas based on bulk $Ga_xIn_{1-x}As$ substrate is that as-grown bulk $Ga_xIn_{1-x}As$ has very low resistivity and is not capable of holding off the high voltages required for the operation of semi-large aperture PC antennas. Attempts to circumvent these problems have resulted in experimental investigation of THz emission from micro-dipole antennas fabricated on low temperature grown $Ga_xIn_{1-x}As$ or heavy ion irradiated thin-films of $Ga_xIn_{1-x}As$, grown on binary substrates by molecular beam epitaxy (MBE). While the ion irradiation process does increase the resistivity of the material to some extent, it reduces the carrier mobility of the material greatly, thus deteriorating the THz emission from the PC antennas based on $Ga_xIn_{1-x}As$ thin films. Further, MBE grown thin films of $Ga_xIn_{1-x}As$ are likely to exhibit high density of misfit dislocations (typically $\sim 10^6$ to $10^9/cm^2$) that originate at the growth interface due to lattice mismatch between the epilayer and substrate [2]. Bulk crystals grown from high temperature melts are free from such defects due to the near-equilibrium growth process and possess the

S. Sengupta, *Characterization of Terahertz Emission from High Resistivity Fe-doped Bulk Ga$_{0.69}$In$_{0.31}$As Based Photoconducting Antennas*, Springer Theses, DOI: 10.1007/978-1-4419-8198-1_6, © Springer Science+Business Media, LLC 2011

highest crystalline perfections. For the work reported in this dissertation, we have collaborated with Professor Partha Dutta at Rensselaer Polytechnic Institute, to successfully grow Fe-doped bulk $Ga_{0.69}In_{0.31}As$ crystals with high resistivity ($\sim 10^7 \Omega$) and high dc Hall mobility (2,395 cm^2/Vs) using a hybrid vertical Bridgman and gradient freezing directional solidification process. Since the THz emission process from both surface emitters and PC antenna emitters is closely associated with the ultrafast carrier dynamics inside the semiconductor material, we have also investigated the same in as-grown $Ga_xIn_{1-x}As$ and Fe-doped $Ga_{0.69}In_{0.31}As$ using ultrafast spectroscopy techniques.

Our experimental results on carrier dynamics in $Ga_{0.69}In_{0.31}As$ and THz emission characteristics of semi-large aperture $Ga_{0.69}In_{0.31}As$ PC antenna establish that this material is a strong candidate for high output power THz emitters driven by multiwatt femtosecond solid state lasers operating at wavelengths of 1–1.5 μm. Along with very high dark resistivity (necessary to hold off high bias fields), bulk $Ga_{0.69}In_{0.31}As$ also possesses sub-picosecond carrier lifetimes and very high values of ultrafast carrier mobilities (about two orders of magnitude higher than that reported in $Ga_xIn_{1-x}As$ thin films) as estimated from our ultrafast experiments and THz emission studies. The sub-picosecond carrier lifetime prevents the buildup of persistent photoconductivity and reduces performance degradation due to thermal runaway, while high ultrafast carrier mobilities favor efficient THz generation from the material. From our experiments we further predict that it is possible to extract an output THz power of 51 μW from a semi-large aperture $Ga_{0.69}In_{0.31}As$ PC antenna with 0.8 mm electrode gap spacing when it is driven at an incident laser power 1 W (obtained from Yb:doped lasers at wavelengths of 1.1 μm), and at an applied bias field of 11.5 kV/cm. This result is indeed promising in that the THz power from a semi-large aperture $Ga_{0.69}In_{0.31}As$ PC antenna is a factor of ten within the range of half a milliwatt of output THz power, and is about two orders of magnitude higher when compared with the output of InAs crystal (a commonly used emitter in most THz time-domain spectroscopy systems).

While these first results are indeed promising, further research needs to be done to optimize the performance of $Ga_{0.69}In_{0.31}As$ PC antennas. In our experiments, the THz output power from such PC antennas was limited by bulk breakdown and surface flashover at an applied bias field of 12.5 kV/cm. The performance of the antennas was also limited by thermal runaway at high optical fluences at high bias fields caused by ohmic heating of the semiconductor substrate, and saturation of the emitted THz power due space charge and near field screening effects. A crack-free semiconductor surface obtained by high quality crystal polishing, and good ohmic contacts formed by large area, polished copper electrodes with rounded edges, can prevent the surface flashover to some extent. Passivation of the semiconductor emitter surface by a suitable dielectric coating can also increase the threshold for bulk breakdown and surface flashover caused by dangling bonds at the semiconductor surface or a surface contamination layer. As demonstrated by Lloyd-Hughes and co-workers [3], the surface passivation of GaAs surface by a thin layer of $(NH_4)_2S$ resulted in the doubling of emitted THz power from a PC antenna formed on the surface.

Substrate heating effects can be minimized by resorting to active cooling mechanisms such as water-cooling and Peltier cooling. Cooling has been reported to produce an overall increase in THz power generated by PC antenna emitters [4]. Saturation of the output THz power due to screening effects can also be reduced by opting emitters with larger electrode gap spacing and homogeneous illumination of the entire antenna gap.

The future recommended research direction, for extending the work reported in this dissertation, would be to characterize the THz emission from $Ga_xIn_{1-x}As$ PC antennas with varying Ga mole fraction. Earlier research by Ko et al. [5] show that the THz emission from $Ga_xIn_{1-x}As$ varies as a function of the Ga mole fraction x, and maximizes for $x \sim 0.4$. Our attempts to grow a high resistive material with Fe doping at this composition was only partially successful in that it increased the resistivity to only about $\sim 100\ \Omega$, which is much lower than the required values for THz PC antenna applications. Nevertheless, it may be possible to develop high resistivity $Ga_xIn_{1-x}As$ crystals at Ga compositions close to $x = 0.69$, and a systematic study should be undertaken in this regards. It would also be insightful to characterize the THz emissions from such high resistivity $Ga_xIn_{1-x}As$ based PC antennas at the emission wavelength of Yb:doped lasers at 1.1 μm and at a higher average laser power of ~ 1 W. We expect that the information obtained from such experimental works will pave the way for the realization of high-power, high-bandwidth, high repetition rate THz emitters in future.

References

1. Reid, M., Fedosejevs, R.: Quantitative comparison of terahertz emission from (100) InAs surfaces and a GaAs large-aperture photoconductive switch at high fluences. Appl. Opt. **44**, 149 (2005)
2. Communication with Professor Partha Dutta, Department of Electrical, Computer and Systems Engineering, Rensselaer Polytechnic Institute, 110 8th street, Troy, NY 12180
3. Lloyd-Hughes, J., Merchant, S.K.E., Fu, L., Tan, H.H., Jagadish, C., Castro-Camus, E., Johnston, M.B.: Influence of surface passivation on ultrafast carrier dynamics and terahertz radiation generation in GaAs. Appl. Phys. Lett. **89**, 232102-1 (2006)
4. Zhao, G., Schouten, R.N., van der Valk, N., Wenckebach, W.T., Planken, P.C.M.: Design and performance of a THz emission and detection setup based on semi-insulating GaAs emitter. Rev. Sci. Instrum. **73**, 1715 (2002)
5. Ko, Y., Sengupta, S., Tomasulo, S., Dutta, P., Wilke, I.: Emission of terahertz-frequency electromagnetic radiation from bulk $Ga_xIn_{1-x}As$ crystals. Phys. Rev. B **78**, 035201 (2008)

Appendix A
Golay Cell Specifications

Golay Cell

SPECIFICATIONS

Diameter of input window..............6 mm
Sensitivity at 12.5 Hz modulation
frequency................................... $1.5 \cdot 10^5$ V/W
Noise equivalent power at 12.5 Hz modulation,
1 Hz bandwidth.............*lower then* 10^{-10} W/Hz$^{1/2}$
Dynamic range...............................10^{-10}... 10^{-5} W
Rise time......................................25 ms
Maximum output voltage....................*0.5*V
Dimensions.................................... $135 \cdot 115 \cdot 120$ mm^3
Weight...1.8 kg
Environmental characteristics:
    Temperature................................+15...+25°C
    Pressure.....................................$8.4 \cdot 10^4$...$10.7 \cdot 10^4$ Pa
    Humidity.....................................45...80%

Golay Cell is a pneumatic detecting cell adapted for the *millimeter-submillimeter* wavelength domain (20...6000μm). The Golay Cell has input window's diameter of 6 mm$_2$ window's materialis polyethylene; the black polyethylene before the window is used additionally to reject optical radiation. The Golay Cell is equipped with build-in preamplifier. The electronics need 3 regulated power supplies +15 V, −15 V and 32 mA for infrared LED.

For operating in *quasioptical* configuration the detector cell is mounted on vibration isolating base which is fixed on the rod holder of the adjustable mounting base, ensuring necessary adjustment of its position relative to an axis of a radiation beam.

Index

MIX
Papier aus verantwortungsvollen Quellen
Paper from responsible sources
FSC® C105338

If you have any concerns about our products,
you can contact us on
ProductSafety@springernature.com

In case Publisher is established outside the EU,
the EU authorized representative is:
Springer Nature Customer Service Center GmbH
Europaplatz 3, 69115 Heidelberg, Germany

Printed by Libri Plureos GmbH
in Hamburg, Germany